'Concise in expression, with handsome cartography and a text
it elucidates the key issues surrounding global climate cha

'Seeing that climate change is probably the biggest single problem facing our world today, it's natural that you might want to find out more. You could wade through dense, dry academic detail from the IPCC. Or you could root out the newly published *Atlas of Climate Change,* which condenses key findings from the scientists and is packed with facts, graphs, and maps explaining how climate change is affecting the planet.' ***Guardian***

'This is a remarkable piece of work and extremely readable. What is heartening is not only the wealth of information it conveys but also the powerful illustrations and graphics.' **Director-General, The Energy and Resources Institute (TERI)**

'This pioneering atlas will become an essential point of reference for anyone looking for a quick and accurate overview of this multidisciplinary subject.' **Foreign and Commonwealth Office, UK**

'An excellent book and wonderful tool to help clarify the mountains of data into something comprehensible. If you want to get up to speed on the subject, this is probably the best place to start. Highly recommended.' ***Transition Culture***

'An excellent vehicle for beginning to understand the climate change issue.' ***Forestry Chronicle,*** **Canada**

'Packs a lot of very high quality information into a small space with an excellent layout. A hit with students and essential for every library shelf – one of the best buys of the year.' ***Ecological and Environmental Education***

'If you want to understand the evidence, causes, and consequences of climate change, as well as what we can do to deal with it, get your copy today.' ***CarbonSense***

'A succinct, yet exacting new resource that will be vital for many years to come.' **Oxford University Centre for the Environment**

'This vast mine of information takes a wider than usual view of the implications of climate change. It is a valuable resource to help understand the core issues and puts into perspective the desperate unfairness of the global picture.' ***Morning Star***

'All schools and colleges should complement their environment studies with this atlas.' ***Education Journal***

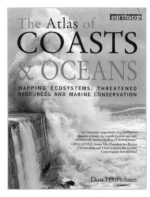

The Atlas of
COASTS
& OCEANS

MAPPING ECOSYSTEMS, THREATENED
RESOURCES AND MARINE CONSERVATION

Don Hinrichsen

In the same series:

'Invaluable...I would not be without the complete set on my own shelves.'
Times Educational Supplement

'No-one wishing to keep a grip on the reality of the world should be without these books.'
International Herald Tribune

'Fascinating and invaluable.' *The Independent*

'A new kind of visual journalism.' *New Scientist*

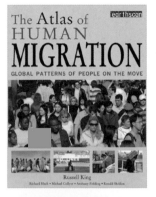

The Atlas of
HUMAN
MIGRATION
GLOBAL PATTERNS OF PEOPLE ON THE MOVE

Russell King
Richard Black • Michael Collyer • Anthony Fielding • Ronald Skeldon

Foreword by
Clive Stafford Smith

The Atlas of
HUMAN
RIGHTS

MAPPING VIOLATIONS OF FREEDOM AROUND THE GLOBE

Andrew Fagan

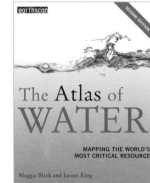

The Atlas of
WATER

MAPPING THE WORLD'S
MOST CRITICAL RESOURCE

Maggie Black and Jannet King

The Atlas of
FOOD

WHO EATS WHAT,
WHERE, AND WHY

Erik Millstone and Tim Lang

The Atlas of
ENDANGERED
SPECIES

Richard Mackay

NEW UPDATED 8TH EDITION
THE
STATE
OF THE
WORLD
ATLAS

THE BESTSELLING SURVEY OF
CURRENT EVENTS & GLOBAL TRENDS

Dan Smith

The Atlas of
WOMEN
IN THE WORLD

Joni Seager

The Atlas of
HEALTH

MAPPING THE CHALLENGES
AND CAUSES OF DISEASE

Diarmuid
O'Donovan

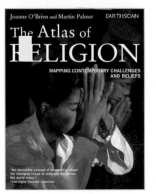

Joanne O'Brien and Martin Palmer   EARTHSCAN

The Atlas of
RELIGION

MAPPING CONTEMPORARY CHALLENGES
AND BELIEFS

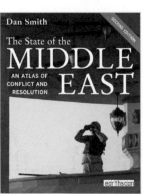

Dan Smith

The State of the
MIDDLE
EAST

AN ATLAS OF
CONFLICT AND
RESOLUTION

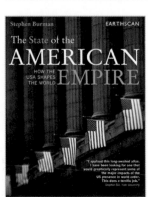

Stephen Burman   EARTHSCAN

The State of the
AMERICAN
EMPIRE

HOW THE
USA SHAPES
THE WORLD

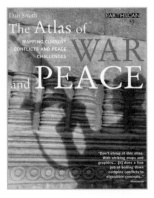

Dan Smith   EARTHSCAN

The Atlas of
WAR
and **PEACE**

MAPPING CURRENT
CONFLICTS AND PEACE
CHALLENGES

The Atlas of CLIMATE CHANGE

Mapping the World's Greatest Challenge

Third Edition

Kirstin Dow and Thomas E Downing

earthscan
publishing for a sustainable future

This third edition published by Earthscan in the UK 2011
First edition 2006; second edition 2007

Copyright © Myriad Editions Limited 2011

A catalogue record for this book is available from the British Library

ISBN: 978-1-84971-217-0

Produced for Earthscan by
Myriad Editions
59 Lansdowne Place
Brighton, BN3 1FL, UK
www.MyriadEditions.com

Earthscan is an imprint of the Taylor & Francis Group, an informa business

Edited and co-ordinated by Jannet King and Candida Lacey
Maps and graphics created by Isabelle Lewis
Design and additional graphics by Corinne Pearlman

Printed and bound in Hong Kong on paper produced from sustainable sources
by Lion production under the supervision of Bob Cassels, The Hanway Press, London

For a full list of publications please contact:

Earthscan
2 Park Square
Milton Park
Abingdon
Oxon OX14 4RN
Web: www.earthscan.co.uk

Earthscan publishes in association with
the International Institute for Environment and Development

Contents

PART 1

PART 2

PART 3

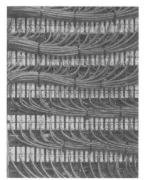

Voices & Visions of Our Future

The *Atlas of Climate Change* should inspire all of us to action. The authors call upon their experience to present the facts on climate change. In a clear format, from the early warning signs to drivers of change, from impacts to policy, they present the weight of evidence. We have come a long way on climate change, from ignorance and denial to policy recommendations and global negotiations.

Over the course of time climate scientists have drawn a line in the sand: the climate is warming and it is projected that unless we change our track we will see a temperature rise of more than 2°C which could have catastrophic effects for the biosphere and all who live in it. Scientists recommend that we should aim at reducing emissions significantly and urgently aspire to become carbon neutral. The scientific evidence so far presented is overwhelming and can be explored in this third edition of the *Atlas*.

The carbon cycle is a key component of ecological systems. And ecological systems are key components of climate action. Deforestation accounts for nearly 20 percent of global carbon emissions, and is reportedly greater than all of the transport systems globally combined. As we all know, the Amazon, Congo Basin, and South-East Asia rainforest ecosystems are the "green lungs" of the planet and are essential for global climate regulation. My work with the women of the Green Belt Movement in Kenya over the past 30 years has shown that grassroots communities will act on the root causes of environmental degradation once they appreciate the linkage between the environment and their livelihoods. It is they who will apply the skills and initiatives that will help them mitigate and adapt against the negative impacts of climate change. It is vital that climate policies work to promote equity, biodiversity, and the rights of vulnerable communities. Solutions to climate change must firmly put people and nature at their core.

We have a moral responsibility to protect the rights of future generations, and of all species that cannot speak for themselves but are nevertheless members of the community of life. The challenge of climate change demands that there be a global political will to address this issue. Without political will, especially of the politically and economically powerful nations, the results will be catastrophic, even as the world continues with diplomatic rhetoric and no action. We are the generation that has the opportunity to effectively respond to this challenge. We are already late. Take action now!

Professor Wangari Maathai
Nobel Peace Prize Laureate, 2004
Founder of the Greenbelt Movement in Kenya, 1977
Goodwill Ambassador to the Congo Basin Forest
UN Messenger for Peace and the Environment
See www.Greenbeltmovement.org for more information on
Professor Maathai's work.

It is not every day that you get a chance to walk on top of an ocean. But that was where I found myself, with five feet of ice the only thing separating me from the 1,000 feet of freezing Arctic Ocean. I was in the Arctic as a correspondent for CNN International, living with scientists working in some of the most extreme conditions on Earth in order to shed more light on the complex systems that drive our climate.

At these temperatures, a human being would survive about five minutes in the water, so it seemed rather counterintuitive to drill and chip through the ice to reach the frigid water below. The US Arctic Research Commission wrote that: "We know more about the topography of the planets Venus and Mars than we do about the bathymetry of the Arctic Ocean." As I stared down at the outlines of the ice hole we had created, it struck me as truly remarkable that such a small window could provide so much new knowledge.

The US federal budget for space exploration is just over 1,000 times larger than that for ocean exploration. While there is no question that exploring the Universe helps us to understand many things about this world, knowing if there is life on Mars is not critical to life on this planet; healthy oceans are. As Arthur C. Clark once wrote, "How inappropriate to call this planet Earth when it is quite clearly Ocean."

As we argue about the status of the environment and what should be done about it, we fail to take action that should be universally embraced: to allow our decisions to be guided by science, and to act in the best interests of future generations. What understanding we do have has led us to a chilling conclusion: that humans are drastically altering the climate, both by emitting huge amounts of CO_2 and by altering the land through agriculture and urbanization in such a way that the stable climate that saw humans thrive for the first millennia of our history is being unbalanced.

According to a report by the World Wildlife Fund we use 1.5 times the amount of resources that Earth can replenish each year. We are like farmers eating our seeds. There is urgency in understanding the world around us, if not for us, than for our children.

My grandfather, Jacques Cousteau, often shared with me a simple dream: that every child born has the right to walk on clean grass under a blue sky, breath fresh air and drink pure water. I believe that his is a dream we all share, and that in order to realize it, we must base our decisions on science, not on wishful thinking.

Philippe Cousteau
Explorer, Social Entrepreneur, Environmentalist and CEO of
EarthEcho International www.earthecho.org

About the Authors

Dr. Kirstin Dow is Professor of Geography at the University of South Carolina. She is the Principal Investigator of the Carolinas Integrated Sciences and Assessments. CISA is part of a NOAA-sponsored network of teams working at the regional level to advance understanding of climate impacts  on society and to improve the value and accessibility of climate science for decision-making. CISA research and collaboration with regional decision-makers also supports the US National Climate Assessment. Her research contributions to this interdisciplinary team focus on climate impacts, vulnerability, adaptation, and communicating uncertainty.

Kirstin's PhD focused on the global to local processes shaping vulnerability for fishing communities along the Straits of Malacca (Geography, Clark University). She is the former Manager of the Poverty and Vulnerability Programme at the Stockholm Environment Institute. Currently, she is a Lead Author in the Intergovernmental Panel on Climate Change fifth assessment, contributing to analysis of adaptation. She also serves on the National Academy of Sciences Board on Atmospheric Sciences and Climate, as an Editor of the American Meteorological Society journal *Weather, Climate, and Society*, and as a science advisor on climate change for community and national efforts. She seeks to inform a transition to a sustainable pathway that recognizes the climate and development challenges in the USA and abroad.

Dr. Thomas E. Downing is President and CEO of the Global Climate Adaptation Partnership (GCAP). GCAP is a new initiative, to provide knowledge-led, climate change adaptation solutions for its clients, training through the Adaptation Academy, and advanced knowledge management services. Tom retains  a Visiting Professor position at Oxford University in the School of Geography and Environment and is an Associate in development studies in Queen Elisabeth House. He was formerly Executive Director of the Stockholm Environment Institute's office in Oxford.

Tom's early career included four years working with the Government of Kenya on environmental assessment, drought, and sustainable development. His PhD was on drought coping strategies in central and eastern Kenya (Geography, Clark University). His major interests are sustainable land and water management, including extensive experience in vulnerability assessment and climate risk management. Involved in all five of the assessments of the Intergovernmental Panel on Climate Change, more recently he has led assessments on the economics of climate change, including adaptation planning, extreme events, and end-use effectiveness for vulnerable populations. His current work on mainstreaming climate adaptation in the African Development Bank highlights practical means for institutional readiness and guidelines for measuring climate protection at the project level. Tom continues to develop practical tools for making sound decisions in order to meet the climate challenges ahead.

Introduction

The year 2010 may well be seen with hindsight as pivotal. It started badly, with the detritus of the Copenhagen Conference of Parties, and chaos around the future of international negotiations. Then, a series of climate extremes rekindled attention. The floods in Australia, China, and Pakistan left unforgettable images of inundation. The Chinese and Russian droughts were of a staggering scale. As we write, another crisis in the Horn of Africa is looming, recalling images of the 1970s famines. There were new entries in the league tables for extreme events, where Bangladesh and the USA dominated headlines. And budget crises continued to remind us of global economic volatility.

Mostly behind the scenes, the international negotiations reconvened, making progress in Cancún. Trust in the process was restored to some extent, if not full confidence in reaching final agreements. By 2011, the Arab Spring captured our imaginations with its realignment of global politics. Attention to the future of the planet is widespread as we near the 20th anniversary of the original Earth Summit in Rio (Rio+20) – another pivotal moment and touchstone for reflection.

In this *Atlas of Climate Change*, each double-page spread introduces central issues in current debates about the physical processes, likely impacts, and adequate responses. We noted in the introduction to the first edition some of our motivation:

> After listening to shrill warnings of catastrophe, dismissive statements from skeptics, decades of calls for more research, incomplete stories of the complexities of negotiations, and confusion over the level of consensus in scientific understanding, we sought to provide a resource to help people make that initial step into understanding the core issues of climate change.

This third edition updates and expands upon the climate story, from increasingly common signs of change and the undisputed basics of the science, through the panoply of drivers and consequences, finishing with the status of international action and personal and public commitments to solutions. Many of the pieces of the puzzle in this story have been known for some time. Increasingly the pieces are illuminating details of the big picture.

In this introduction we reflect on what we have seen and learned since 2005, when we began the first edition. There have been substantial changes surrounding the understanding and treatment of climate change. As we collated the evidence, again, reviewed major points, and reflected, we are personally deeply concerned. Uncertainty remains, but it is absolutely clear that

forestalling action on that basis is to invite other problems and losses. Uncertainty is no longer an excuse for inaction. These are our key messages, sources of inspiration and challenges for the times ahead.

Almost all governments have committed to action on climate change

Signatories to the UN Framework Convention on Climate Change, most of the countries in the world, have committed to preventing "dangerous anthropogenic [human-induced] climate change", the key objective of the Convention. In Cancún, countries went further in marking their commitments. The Conference of Parties confirmed that "climate change is one of the greatest challenges of our time." The Cancún Agreements (registered as decision 1/CP.16, or FCCC/ CP/2010/7/ Add.1) further affirmed:

- Scaled-up overall mitigation efforts that allow for the achievement of desired stabilization levels are necessary, with developed country Parties showing leadership by undertaking ambitious emission reductions and providing technology, capacity-building and financial resources to developing country Parties…
- Adaptation must be addressed with the same priority as mitigation…
- Recognizes that warming of the climate system is unequivocal and that most of the observed increase in global average temperatures since the mid-twentieth century is very likely due to the observed increase in anthropogenic greenhouse gas concentrations…
- Further recognizes that deep cuts in global greenhouse gas emissions are required according to science…with a view to reducing global greenhouse gas emissions so as to hold the increase in global average temperature below 2°C above preindustrial levels…

These are strong statements of commitment which are firmly rooted in a vast knowledge, touching every aspect of science and life. This understanding does not eliminate uncertainty altogether; but amid the complexity, uniqueness, and variability, there are topics of consensus, supported by large bodies of scientific literature.

The pace of change is accelerating, in an increasingly volatile world

The rate of climate change is unprecedented in the history of modern human societies, certainly since the Industrial Revolution. Arguments over much longer histories are important for the science of climate change, but hardly comforting when we consider the state of the world today. There are indications that global emissions are returning to a high-growth scenario, one that was considered almost unimaginable just a decade ago.

The broad-reaching effects of climate change are becoming more apparent, and not just to scientists but to people in their daily lives. Residents in parts of Alaska have already experienced warming of up to 4°C. They have seen buildings subside as the ground beneath them thawed for the first time in memory. Old photographs retain the images of glaciers, seemingly invulnerable, that are now gone. Around the world, the predictions of climate change – that we would much rather dismiss – are now all too obvious.

The backdrop of climate policy has changed dramatically in the past five years. Decades of economic growth and stability (for some) have been overturned with global recession, shrinking budgets and volatile markets. As we write, European governments are contemplating a second $100 billion bailout of Greece in hopes of stabilizing the Euro-zone and avoiding a cascade of defaults that could capture Ireland, Portugal, Spain, and even Italy. While commitments for development assistance and climate change are by and large protected, no one believes major new and additional funding for climate change will be easy to mobilize. Financial crisis has clear impacts on real people, across the world. The vision of our connected world captured at the Rio Earth Summit in 1992 seems more dangerous.

In this edition, we include a new spread based on the limits of prediction and responses. The conjuncture of shocks in biogeophysical processes (often referred to as tipping elements), fragility in governance and large-scale impacts could have alarming consequences. For instance, collapse of the West Antarctic Ice Sheet and deglaciation of Greenland would raise sea levels by some 12 meters. Retrenchment of poverty and civil strife in poor countries would swamp nascent efforts to avoid additional climate impacts. Population displacement due to climate-related disasters, sea-level rise, and collapse of regional production systems destabilizes governments and accelerates vulnerability.

Climate as an issue of security has gained the attention of several forums. Not all climate-related changes will be negative! The Arab Spring, in some ways related to high food prices and the impacts of climate extremes, caught us by surprise. Greater awareness of climate change is leading to new opportunities for development too.

Action is gaining ground, but is hardly adequate

The changes already in evidence severely challenge environmental management and international governance. As we approach Rio +20, we still lack effective agreements to stabilize the level of greenhouse gases in the atmosphere and to protect the most vulnerable populations. Commitments under the Kyoto Protocol and the more recent round of voluntary targets are only the beginning of the 60 percent to 80 percent reductions required.

Dependence on fossil fuels is deeply embedded in our economies and political regimes. Higher oil prices have sparked a boom in shale oil, and pushed forward more hazardous technologies. It seems we are going to have to make decisions to limit fossil-fuel use rather than rely on prices or peak oil to achieve drastic reductions. At the same time, the Deepwater Horizon and Fukiyama disasters have pushed many people and some governments to promote renewable energy.

Globalization, with its impacts on industrial location, economic development, and extensive transport demands, is influencing future patterns of emissions. Negotiating reductions is complicated by the combination of historical versus anticipated contributions associated with economic development, and export relations that result in one country producing goods and emissions for markets in another.

Climate change is as much a humanitarian and human development concern as it is an environmental one. The most vulnerable populations are not the people driving climate change. Of those living on less than a dollar a day, few have electricity, cars, refrigerators, or water heaters. But, because their lives are tied to climate conditions and they have few resources to buffer against bad or progressively difficult conditions, they are likely to bear the highest human costs. It is encouraging to see approaches based on human rights continue to be influential in negotiations.

Connections are also being made between adaptation and mitigation. The cost of responding to disasters may mean cutting other budgets. New Orleans cut funding for education, police, social, legal, and other services after Hurricane Katrina. In the UK, assistance for peacekeeping and humanitarian emergencies is 10 percent of official development assistance. Even wealthy, well prepared countries will face trade-offs between action at home and abroad, between risk reduction and emergency responses, and possibly between adaptation and mitigation.

Responding to climate change and the threatened consequences is a serious and sobering business, but it is not hopeless, nor are all foreseeable outcomes inevitable. There are feasible, effective solutions. Many of these actions save money and make good sense for other reasons as they also address the broader issues of sustainability through reducing dependency on increasingly scarce petroleum supplies, cutting air pollution, and preserving the ecosystems that help maintain the ecological processes we rely on. Although it is tempting to focus on technological solutions, it is essential also to address the issues of consumption and lifestyle that drive technologies.

Technological change alone cannot achieve the necessary levels of emissions reductions; many currently available technologies can, however, make a substantial difference. Renewable energy technologies are becoming more diverse, affordable, and widely adopted. These technologies offer great potential for greenhouse gas reduction as they currently account for less than 5 percent of global energy production. Greater efficiencies are possible in other manufacturing and power production systems as well. Interest in nuclear power has also risen, but its future role in addressing climate change issues is less certain. Concerns over safety and long-term storage of radioactive wastes remain, and it is not clear that its potential value as a response to climate change offers sufficiently strong justification to overcome other economic barriers.

We have moved beyond the notion that installing energy-saving lightbulbs is all it takes: every bit helps, but all the bits need to be in play. Achieving needed reductions will involve adjustments to infrastructure and institutional practices, such as new building designs and training of architects and engineers. Yet, without changes in consumer demand, such as reducing the desire for ever larger homes and more powerful cars, these advances and opportunities could be fruitless.

The necessary institutional innovations also depend on finance. Carbon trading is growing: markets such as the EU trading scheme and the Regional Greenhouse Gas Initiative in the USA are fully functioning. The level of trade is growing rapidly, but still represents a very small percentage of world carbon emissions. Pricing carbon in forests and terrestrial stores, as proposed in the REDD+ initiative, could transform local communities. Caution is warranted to ensure finance is secure and benefits those most dependent on forest resources. International parlance calls this the issue of leakage; empowerment, transparency and accountability are stronger watch-words to ensure corruption and control by the state does not overwhelm the global environmental benefits.

Action will take new institutional strategies and forms of cooperation, and a willingness to deal with longer time frames in decision making. As scientists who for years have been studying the consequences of climate and other environmental changes on people and livelihoods, we believe climate change and its potential impacts are extremely serious issues. The required changes will involve us all deeply, but that social vision of a path to the future is not fully articulated. We seem to bounce between divided notions of fragmented communities preserving their own ideals and lifestyles and a global, shared vision of collective responsibility and action.

Adaptation is an imperative, not a substitute for mitigation

The commitment to warming is already in place – nothing we do now will alter global warming until the 2040s at best. Another two generations will be born, and the global population is likely to have added 2 billion people before our actions take effect. While the legacy of inaction and latent forcing of accumulated greenhouse gases are well-known features of climate change, the imperative to adapt with foresight and reduce the risks associated with the climate commitment has gained momentum only in the past decade or so.

An odd feature of international negotiations has been that environmental campaigners resisted any mention of adaptation as a solution, fearing it would be seen as an alternative to mitigation and let the world off the hook of reducing emissions. This view restricted efforts to develop international adaptation funds and programmes. With growing understanding of the commitments and possible impacts, the IPCC fourth assessment clearly stated that adaptation and mitigation are policy complements, that is, they must both be pursued, and one does not substitute for inaction on the other.

People, economies, and ecosystems are at serious risk from current climate patterns. Across large parts of Africa and Asia the timing and the abundance of rains determines whether crops will support households, or whether hungry people will need to search for alternative sources of food and income. Some coastal shorelines have limited protection from storms. Events of the past decade have reminded us that even the most advanced industrialized countries have an adaptation deficit, a lack of preparedness in the ability to deal with the present, much less the future.

Those at risk are not randomly distributed; nor is risk accidental. Looking ahead, while there will be benefits for some, there will be severe consequences for others. Poverty is an enduring feature of how we organize our economies and societies. Those presently most vulnerable – poor people in marginal areas – are likely to suffer first. Nearly one person in four in developing countries lives in poverty, on less than $1.25 per day. These people often depend heavily on agriculture, fishing, and animal husbandry, or work in the urban informal economy in precarious housing, to maintain their livelihoods and hopes for better lives. With 2°C warming, millions more people, many of them among the poorest, are likely to be exposed to annual coastal flooding. Changing precipitation patterns, either wetter or drier, along with altered temperature, will affect crop productivity and availability of food and water for livestock. Climate changes may also facilitate the movement of human, plant, and animal diseases into areas where they were previously little known, and where doctors, veterinarians, agriculture extension specialists, and money for treatments are all in short supply.

Not all serious climate impacts will be the direct result of local changes. The interconnectedness of the global economy is a major mechanism for transferring impacts from one region to another and among people. For instance, favorable growing conditions worldwide and a bumper crop might translate into lower prices for all producers, while a crop loss for one area may create an economic windfall for another.

The risks are pervasive, and also critical in the distribution and functioning of ecosystems. Many ecosystems are currently under tremendous pressure from land-use change and over-harvesting. Climate change is emerging as another unavoidable stress that will require action to support adaptation to multiple stresses.

A broad front of adaptation is being pursued. Community-based adaptation focuses on livelihoods and local action. Sectoral actions look to promote climate resilience. National plans set policy frameworks, monitor outcomes and support finance. Social networking and information platforms are now common. Concerted urban adaptation programmes have gained momentum in the past five years.

However, there are limits to adaptation. The ability of entire ecosystems to shift is uncertain, both because the rate of change is faster than typical ecological timescales, and because there is limited space available to move. The area immediately surrounding protected areas and coastal areas is often already in use and, in the far north and on mountain tops, there are no more options. Social, institutional, and economic barriers to restructuring infrastructure and economic wellbeing may well be insurmountable. Imagine trying to rebuild a city like Mumbai to cope with the plethora of climate risks it already experiences. And to do that for every coastal city.

Climate change touches everything, and everyone
The breadth of climate change is enormous. From the complexity of the science to the dimensions of impacts, from the structures of vulnerability to the transformations required in economic relationships. The lens of climate change reveals much about the interconnectedness of our world. The many dimensions of complexity and the difficulties they pose for generating solutions makes climate change what some academics have identified as a wicked environmental problem. These wicked traits are part of the way climate change both creates and requires new types of connections among us and future generations:

- The solution depends on how the problem is framed – no single solution can solve the full breadth of climate change, so we have to break it down into parts that can be addressed. Witness the negotiations to appreciate the herculean task of creating pieces that can garner agreement.
- Stakeholders have radically different frames for understanding the problem – whether formal regimes or cultural personalities, we describe the causes and solutions to climate change quite differently. Some see it as a problem of the global commons, others frame

solutions through economic growth with new technology.
- There is no central authority – the UNFCCC is a coordinating framework but cannot impose solutions.
- The long term is valued more than expected – scenarios of policy effects and residual impacts are routinely run out to 2100, and some explore even longer horizons.
- The problem is never solved definitively – actions we take today will not make the climate challenges disappear. We are already committed to impacts for at least another 100 years.
- Those who are trying to solve the problem are also the cause – the imbalance between industrial emissions over the past century (largely North America and Europe) and the next century (with the BASIC countries taking a large share) plagues the negotiations; or more personally, wealthy consumers are also the political elite.
- Time is running out – how long we have before the climate challenge becomes unsolvable is a critical question, and not just academic.

Solutions will require "all hands on deck". All levels of effort, from individuals to cities, to nations, corporations, and the international community are needed to meet this challenge. From individuals taking part in public demonstrations such as the 350 movement, to business councils offering toolkits for reductions, to strong national programs and active international negotiations, this engagement is taking place. An encouraging sign is the rising role of social entrepreneurs. New leadership is coming forward at all levels, personally committed to taking action. New initiatives, companies, and web services are launched every day.

Action on climate change is a part of diverse global and local movements. It is grounded in environmental science and action, and in concerns for sustainability. It is often linked to other environmental issues, such as ozone depletion, or economic trends, such as the peak in oil production. Our experience of climate change is intimately exposed in the record of natural disasters. And disaster reduction has a humanitarian urgency as we seemingly lurch from crisis to crisis.

We have much to learn. Learning pathways need to be constructed, scaling up success. Such pathways are journeys through the challenging landscape in three particular ways.

First, is the action of each of us. There is much we can do in our own lives and within the remits of our own organizations. Promoting a positive psychology of action is essential.

The second level of a journey often comes close upon our individual spheres of action. At some point we must involve others. This may be as simple as achieving a scale of action to warrant investment in a new technology – the market imperatives. Cooperation is also central to designing "fair" solutions, between industrialized nations and least developed countries, or in limiting luxuries such as very large cars while protecting the poor from high transport costs.

Cooperation alone is not sufficient: the third level is transformation. This is the more difficult landscape, where new organizations and institutions are required. It may involve replacing entrenched gatekeepers who are preserving their narrow self-interests, while seemingly supporting action on climate change. The ultimate solutions to the challenges ahead require action far beyond adding a bit of sustainability to what we are doing today.

The transformative pathways begin in concern for environment (the historic locus of climate change), disaster risk reduction (mostly to cope with climate impacts, but increasingly concerned about planetary futures), and economic investment (from development horizons to technological innovation). The goals of sustainable environments, saving lives and livelihoods, and ending poverty are guiding lights. We believe solutions to the challenges and opportunities of climate change must reflect each of these goals.

We encourage you to get involved, in your own lives, in your own, even virtual, communities, and in collective endeavors.

Kirstin Dow
South Carolina, USA
Tom Downing
Oxford, UK
June 2011

Acknowledgements

Many people have shaped our efforts, here in the Atlas and beyond. We would like to pay homage to champions who have passed away since the first edition of the Atlas. Steve Schneider led the world in many scientific ways and as a champion of informed policy. Bo Lim worked tirelessly to do adaptation, from the first Adaptation Policy Framework to projects around the world. Gilbert White inspired many in pioneering work on sustainability and disasters. We miss them in our lives and in the stage of climate policy.

Scientists, interns and assistants around the world have helped on various topics, and this edition builds on the previous versions. Particular contributions to the third edition are noted from Andrew Alberico, Lisa Alexander, Sally Brown, Sujatha Byravan, Greg Carbone, Mark de Blois, Lesley Downing, Kristie Ebi, Mo Hamza, Tomotaka Iba, Sari Kovats, David Lobell, Elizabeth Marino, Karly Miller, Robert Nicholls, David Nimitz Steve Pye, Chella Rajan, David Stainforth, Nassos Vafeidis, Paul Watkiss. And many others who gave us advice, answered emails or made lasting contributions to the data, science and assessments.

Many of the websites that provide data have been vastly improved over the past five years. Those behind the scenes of making data reliable and available are often unsung heroes of climate change; they are deeply appreciated.

The team at Myriad Editions brought the data to life: Jannet King, Candida Lacey, Isabelle Lewis, Corinne Pearlman. Jonathan Sinclair Wilson, formerly at Earthscan, continues to provide a sharp insight into public understanding of science. A special thanks to Wangari Maathai and Philippe Cousteau. Their forewords speak for themselves; their lives have touched us deeply.

15

Definition of Key Terms

Definitions for chemical names, units, technical terms, and regional groupings recognized in international treaties are provided below, along with explanatory notes on impacts, vulnerability and adaptation, theory, predictions, forecasts and scenarios, and the IPCC suite of scenarios. Sources for the definitions are provided at the end of the book.

Chemical names

CCl₄ Carbon tetrachloride.

CFC Chlorofluorocarbon – covered under the 1987 *Montreal Protocol on Substances that Deplete the Ozone Layer* and used for refrigeration, air conditioning, packaging, insulation, solvents, or aerosol propellants. Since they are not destroyed in the lower **atmosphere**, CFCs drift into the upper atmosphere where, given suitable conditions, they break down **ozone**. These gases are being replaced by other compounds, including **HCFCs** and **HFCs**, which are greenhouse gases covered under the **Kyoto Protocol**.

CH₄ Methane.

CO₂ Carbon dioxide.

CO₂e Carbon dioxide equivalent. *See under* Technical terms.

Halocarbons Compounds containing carbon and one or more of the three halogens: fluorine, chlorine, and bromine, including the greenhouse gases CFCs, CCl₄, and HFCs.

HCFC Hydrochlorofluorocarbon.

HFC Hydrofluorocarbon.

N₂O Nitrous oxide.

O₃ Ozone in the lower atmosphere (troposphere) that acts as a **greenhouse gas**. It is created both naturally and by photochemical reactions involving gases resulting from human activities ("smog"). In the stratosphere, ozone is created by the interaction between solar ultraviolet radiation and molecular oxygen (O_2). In the upper atmosphere (stratosphere) ozone plays a decisive role in the stratospheric radiative balance. Its concentration is highest in the ozone layer. Depletion of stratospheric ozone results in increased ultraviolet radiation.

Units

GW Gigawatt.

GWth Gigawatt thermal.

micron One millionth of a meter.

Tonnes Metric tons, equivalent to 1,000 kg or 2,204.62 lb. A gigatonne is one billion (10^9) tonnes.

Technical terms

Anaerobic A life or process that occurs in, or is not destroyed by, the absence of oxygen.

Anthropogenic Resulting from or produced by human beings.

Calving The breaking away of a mass of ice from a floating glacier, ice front, or iceberg.

Carbon dioxide equivalent (CO₂e) A measure used to compare the emissions from various greenhouse gases, based on their global warming potential (GWP). The carbon dioxide equivalent for a gas is derived by multiplying the tonnes of the gas by the associated GWP.

Carbon equivalent A measure used to compare the emissions of different greenhouse gases, based on their global warming potential (GWP). Convert from CO_2 to carbon by multiplying by 12/44.

Carbon sequestration The removal and storage of carbon from the atmosphere in **carbon sinks**.

Carbon sink Reservoirs for carbon, such as forests and oceans, processes, activity or mechanisms that store more carbon than they release.

CDM Defined in Article 12 of the Kyoto Protocol, the Clean Development Mechanism is intended to meet two objectives: (1) to assist Parties not included in Annex I in achieving sustainable development and in contributing to the ultimate objective of the convention; and (2) to assist Parties included in Annex I in achieving compliance with their quantified emission limitation and reduction commitments.

Climate In a narrow sense this is usually defined as the "average weather" or the statistical description of the mean and variability of relevant quantities over a period of time ranging from months to thousands or millions of years. The classical period is 30 years, as defined by the World Meteorological Organization (WMO). The relevant quantities are most often surface variables such as temperature and precipitation. Climate in a wider sense is the state of the climate system.

Climate change A statistically significant variation in either the mean state of the **climate** or in its variability, persisting for an extended period (typically decades or longer). Climate change may be due to natural internal processes or external **radiative forcing**, or to persistent **anthropogenic** changes in the composition of the atmosphere or in land use. The **UNFCCC**, in its Article 1, defines it as: "a change of climate which is attributed directly or indirectly to human activity that alters the composition of the global atmosphere and which is in addition to natural climate variability observed over comparable time periods." This Atlas generally follows the UNFCCC's distinction between "climate change" attributable to human activities altering the atmospheric composition, and "climate variability" attributable to natural causes. Although often used to mean climate change, global warming is only one aspect of this – the increase in global mean temperature.

COP Conference of Parties to the UNFCCC.

Coral bleaching The paling in color of corals resulting from a loss of symbiotic algae, in response to abrupt changes in temperature, salinity, and turbidity.

Cryosphere Component of the climate system consisting of all snow, ice, and permafrost on and beneath the surface of the earth and ocean.

Ecosystem A system of interacting living organisms together with their physical environment, which can range from very small areas to the entire Earth.

El Niño/Southern Oscillation A climate pattern characterized by variations in the temperature of the surface of the tropical eastern Pacific Ocean (warming or cooling, known as El Niño and La Niña respectively) and air surface pressure in the tropical western Pacific (the Southern Oscillation). Occurs roughly every five years.

Emissions Reduction Units Equal to 1 **tonne** of carbon dioxide emissions reduced or sequestered arising from a Joint Implementation (defined in Article 6 of the Kyoto Protocol) project calculated using the **global warming potential (GWP).**

Fugitive emissions Intentional or unintentional releases of gases from anthropogenic activities such as the processing, transmission or transportation of gas or petroleum.

Geothermal Literal meaning: "Earth" plus "heat". To produce electric power from geothermal resources, heat from underground sources is tapped by wells and the steam fed through turbines.

Glacier A mass of land ice flowing downhill. A glacier is maintained by accumulation of snow at high altitudes, balanced by melting at low altitudes or discharge into the sea.

Global warming Increase in global mean temperature.

Global warming potential (GWP) An index, describing the radiative characteristics of well-mixed greenhouse gases, that represents the combined effect of the differing times these gases remain in the atmosphere and their relative effectiveness in absorbing outgoing long-wave radiation. The GWP of carbon dioxide is 1.

Greenhouse gas Gases in the atmosphere, both natural and anthropogenic, that absorb and emit radiation at specific wavelengths within the spectrum of long-wave radiation emitted by the Earth's surface, the atmosphere, and clouds. This property causes the greenhouse effect. Water vapor (H_2O), carbon dioxide (CO_2), nitrous oxide (N_2O), methane (CH_4), and ozone (O_3) are the primary greenhouse gases in the Earth's atmosphere, but there are a number of entirely human-made greenhouse gases, such as the halocarbons and other chlorine- and bromine-containing substances, dealt with under the Montreal Protocol.

Gridded data The result of converting scattered individual data points into a regular grid of calculated, hypothetical values. Also known as "raster data".

Ice sheet A mass of land ice that is sufficiently deep to cover most of the underlying bedrock topography. There are only two large ice sheets in the modern world, on Greenland and Antarctica.

Ice shelf A floating ice sheet of considerable thickness attached to a coast (usually of great horizontal extent with a level or gently undulating surface); often a seaward extension of ice sheets.

ICT Information and communications technology.

IPCC Intergovernmental Panel on Climate Change. Established by the World Meteorological Organization (WMO) and the United Nations Environment Programme (UNEP) in 1988 to assess scientific, technical and socio-economic information relevant to the understanding of climate change, its potential impacts and options for adaptation and mitigation. The IPCC Fifth Assessment Report (AR5) is scheduled for release in 2013/2014, based on the results of three working groups involved in assessing the scientific basis, the impacts, adaptations and vulnerabilities, and the mitigation of climate change.

JI Joint Implementation, a market-based implementation mechanism defined in Article 6 of the Kyoto Protocol, allowing Annex I countries or companies from these countries to implement projects jointly that limit or reduce emissions, or enhance sinks, and to share the Emissions Reduction Units. JI activity is also permitted in Article 4.2(a) of the UNFCCC.

Kyoto Protocol to the United Nations Framework Convention on Climate Change (UNFCCC), adopted at the Third Session of the Conference of the Parties to the UNFCCC in 1997 in Kyoto, Japan. It contains legally binding commitments, in addition to those included in the UNFCCC. Countries included in Annex B of the Protocol (most countries in the Organisation for Economic Cooperation and Development, and countries with economies in transition) agreed to reduce their anthropogenic **greenhouse gas emissions** by at least 5% below 1990 levels in the commitment period 2008 to 2012. The Kyoto Protocol entered into force in 2005 after Annex I countries representing at least 55 percent of emissions by industrialized countries ratified it.

Long-wave radiation Radiation emitted by the Earth's surface, the atmosphere, and clouds, also known as terrestrial or infrared radiation.

Mitigation Anthropogenic intervention to reduce the sources or enhance the sinks of greenhouse gases.

MOP Meeting of Parties to the Kyoto Protocol.

Multidecadal variability Climate variations occurring over timescales of one to three decades.

Paleoclimate Climate for periods prior to the development of measuring instruments, including historic and geologic time, for which only proxy climate records are available.

Photovoltaics Panels used to convert the sun's radiation into electricity, also called solar cells.

Precipitation Water in solid or liquid form that falls to Earth's surface from clouds.

Proxy climate indicator A preserved record of climate conditions before the instrumental record, derived using physical or biophysical principles.

Radiative forcing Change in the net vertical radiation at the boundary between the lower and upper atmosphere (the tropopause) due to an internal change or a change in the external forcing of the climate system, such as a change in the concentration of carbon dioxide or the output of the sun.

Terrestrial Relating to the land.

Thermohaline circulation Large-scale density-driven circulation in the ocean, caused by differences in temperature (thermo) and salinity (haline).

UNFCCC United Nations Framework Convention on Climate Change, adopted on 9 May 1992 in New York and signed at the 1992 Earth Summit in Rio de Janeiro by more than 150 countries and the European Community. Its ultimate objective is the "stabilization of greenhouse gas concentrations in the atmosphere at a level that would prevent dangerous anthropogenic interference with the climate system." It contains commitments for all Parties and entered into force in March 1994.

Weather The state of the atmosphere at some place and time described in terms of such variables as temperature, cloudiness, precipitation and wind.

Regions

Annex B countries are listed in Annex B in the **Kyoto Protocol** and have agreed to a target for their greenhouse gas emissions. They include the **Annex I countries** except Turkey and Belarus.

Annex I countries Group of countries included in Annex I (as amended in 1998) to the **UNFCCC**, including all the developed countries in the **OECD**, and economies in transition. By default, the other countries are referred to as non-Annex I countries. Under Articles 4.2(a) and 4.2(b) of the Convention, Annex I countries commit themselves to the aim of returning to their 1990 levels of greenhouse gas emissions by the year 2000.

EIT Economies in transition, countries with national economies in the process of changing from a planned economic system to a market economy. Refers to the former communist countries of Europe.

EU European Union. Data for 2000 to 2004 refer to 15 members: Austria, Belgium, Denmark, Finland, France, Germany, Greece, Ireland, Italy, Luxembourg, Netherlands Portugal, Spain, Sweden and United Kingdom. Data for 2004 onwards refer to 25 member countries, which includes Cyprus, the Czech Republic, Estonia, Hungary, Latvia, Lithuania, Malta, Poland, Slovakia and Slovenia.

G8 Group of Eight countries: Canada, France, Germany, Italy, Japan, the UK, USA, and Russia. The heads of government hold an annual economic and political summit meeting in the country currently holding the rotating presidency.

LDCs The least developed countries, identified as such by the United Nations.

OECD Organisation for Economic Co-operation and Development. 30 member countries sharing a commitment to democratic government and the market economy.

OPEC Organization of the Petroleum Exporting Countries.

Impacts, vulnerability, and adaptation

Climate change impacts are the consequences of natural and human systems. The impacts depend on the **vulnerability** of the system, in the climate change literature defined as a function of the character, magnitude, and rate of climate variation to which a system is exposed, its sensitivity, and its adaptive capacity. However, vulnerability has other common definitions. In disaster planning, risk is the outcome of vulnerability (social, economic and environmental exposure and sensitivity) and hazard (the probability and magnitude of an extreme event). In development planning and poverty assessment, vulnerability is described as exposure to multiple stresses, to shocks and to risk over a longer time period, with a sense of defencelessness and insecurity.

Both impacts and vulnerability may be reduced by **adaptation** – adjustments in natural or human systems to a new or changing environment. Various types of adaptation can be distinguished, including anticipatory and reactive adaptation, private and public adaptation, and autonomous and planned adaptation. For people, adaptation can be seen as a process of social learning. **Adaptive capacity** is the ability to understand climate changes and hazards, to evaluate their consequences for vulnerable peoples, places and economies, and to moderate potential damages to take advantage of opportunities, or to cope with the consequences.

Theory, prediction, forecasts, and scenarios

A scientific **theory** is a coherent understanding of some aspect of our world, based on a well-established body of observations and interpretation. In popular culture, the label "theory" may be used in a derogatory manner to refer to propositions put forward to challenge a mainstream view.

There is some confusion regarding how we view the future. In experimental language, we talk of predicted outcomes. So, a computer simulation of climate change predicts global warming, say of 3°C by 2100. This prediction depends on a set of underlying assumptions.

A climate *prediction* is usually in the form of probabilities of climate variables such as **temperature** or **precipitation**, with lead times up to several seasons. The term climate *projection* is commonly used for longer-range predictions that have a higher degree of uncertainty and a lesser degree of specificity. For example, this term is often used to describe future **climate change**, which depends on the uncertain consequences of greenhouse gas emissions and land use change, in addition to the feedbacks within the atmosphere, oceans and land surface.

It is extremely difficult to anticipate future greenhouse gas emissions. Countries may adopt stronger controls, industry and technology might reduce their carbon intensity, or consumers might rebel and ecological feedbacks might accelerate climate change. Where our understanding of the future is weak, we often use the term **scenario:** a plausible and often simplified description of how the future may develop, based on a coherent and internally consistent set of assumptions about key driving forces and relationships. Scenarios are neither predictions nor forecasts and sometimes may be based on a narrative storyline.

Greenhouse gas emission scenarios

To enable comparisons to be made between scenarios, in 1996 the IPCC issued its *Special Report on Emissions Scenarios* (SRES), which outlined a number of possible scenarios. These are now used by scientists to clarify the assumptions behind different emission pathways and the consequences for climate change.

Scenario A1 represents a future world of very rapid economic growth, low population growth, and rapid introduction of more efficient technologies. Major underlying themes are economic and cultural convergence and capacity building, with a substantial reduction in regional differences in per capita income. The A1 scenario family develops into three alternative directions of technological change in the energy system: fossil-intensive (A1FI), non-fossil energy sources (A1T), or a balance across all sources (A1B).

Scenario A2 portrays a very heterogeneous world. The underlying theme is that of strengthening regional cultural identities, with high population growth rates, and less concern for rapid economic development.

Scenario B1 represents a convergent world with a global population that peaks in mid-century, rapid change in economic structures toward a service and information economy, with reductions in material intensity, and the introduction of clean and resource-efficient technologies. The emphasis is on global solutions to economic, social, and environmental sustainability, including improved equity, but without additional climate initiatives.

Scenario B2 depicts a world in which the emphasis is on local solutions to economic, social, and environmental sustainability. It is a heterogeneous world with less rapid, and more diverse technological change than in A1 and B1.

In 2010, the IPCC developed a scenario framework based on common profiles of greenhouse gas emissions, called Representative Concentration Pathways (RCPs). Climate results using these emissions scenarios will form the basis for the IPCC AR5.

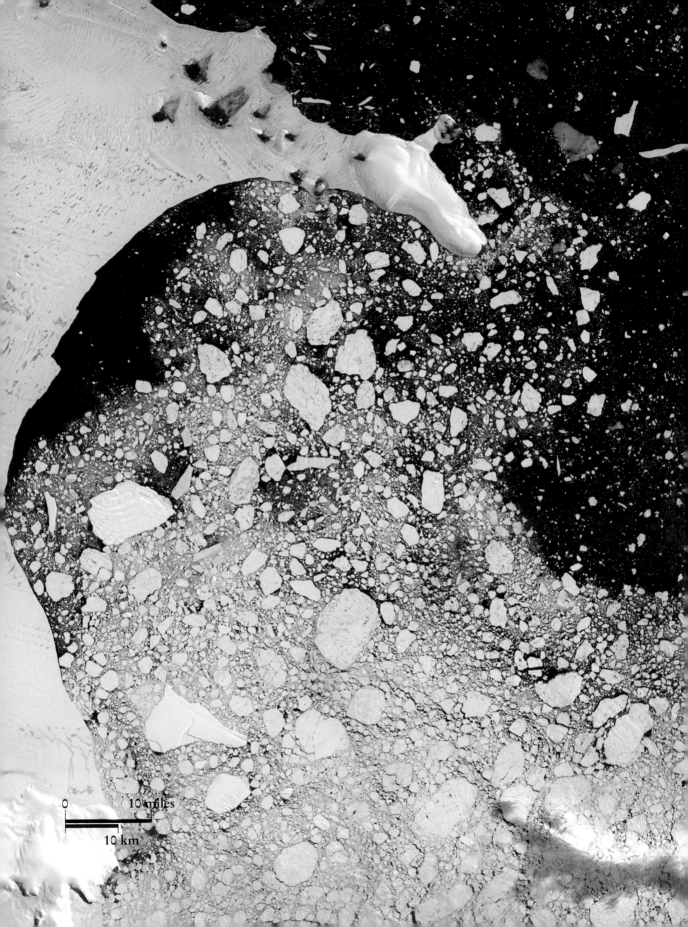

PART 1 Signs of Change

In 2010 and 2011, floods in Pakistan, Australia, and China; heat waves and forest fires in Russia and in the USA; drought in the Amazon, and record-breaking temperatures around the world illustrated that the climate is already dangerous. The global average temperature in 2010 tied for the warmest year on record. The minimum extent of Arctic sea ice was the third-lowest measured. This first decade of this new millennium was itself the warmest observed.

While there are uncertain elements in our knowledge of climate change, and this knowledge is sketchier in some areas than others, the big picture is becoming increasingly clear. The issue is also becoming increasingly urgent. Many of the record-breaking events were accompanied by vast human tragedies.

The science that underpins the big picture draws on tens of thousands of data sets and millions of individual observations. These data track a diversity of physical and biological indicators such as the timing of budburst and flowering in plants and trees, of changes to ice melt on rivers and lakes, and of alterations in the ranges of mammals, birds, and insects. On mountains and at the poles, glaciers are thinning and retreating, ice sheets are breaking up. In many parts of the world, shifting patterns of rainfall intensity and of temperature are affecting people's lives and livelihoods. The oceans are warming, and the increased concentration of carbon dioxide in the atmosphere is driving an increase in ocean acidity, threatening corals and small organisms at the base of the food chain, and, therefore, the survival of entire ecosystems.

Of course, many uncertainties limit our current understanding of the rate, magnitude, and patterns of change. But, despite the uncertainties, the era of human-induced climate-change impacts has begun. The world has moved from warning signs and hints of climate change, to monitoring the increasing scale of impacts and bearing the consequences.

Climate change has a taste, it tastes of salt.

Atiq Rahman
UNEP Champion of the Earth 2008, Director, Bangladesh Centre for Advanced Studies

RECORD HIGHS

National temperature records set in 2010

°C

Pakistan	53.5
Kuwait	52.6
Iraq	52.0
Saudi Arabia	
Qatar	50.4
Sudan	49.7
Niger	48.2
Chad	47.6
Burma	47.2
Bolivia	46.7
Cyprus	46.6
Russia	45.4
Zambia	42.4
Colombia	42.2
Ukraine	42.0
Belarus	38.9
Finland	37.2
Solomon Islands	36.1
Ascension Islands	34.9

Among the thousands of warning signs of climate change, the array of extreme events that took place in 2010 stand out.

Current climate change is affecting all continents and most oceans. Thousands of case studies of physical changes (such as reduced snow cover and ice melt) and changes in biological systems (such as earlier flowering dates and altered species distributions) have correlated with observed climate changes over the past three decades and more. Scientists have high confidence that these environmental changes are part of the early warning signs of climate change.

Effects on social and economic activities are harder to attribute to climate impacts, although major events attract considerable attention. From prolonged drought in Africa and Australia to the dire flooding in Australia, China, and Pakistan, livelihoods, economies, and politics are at risk.

A single extreme weather event or change in the natural environment does not prove that humans are changing the climate. However, the proven physical science, the history of recent observations, and the consistency in model assessments all support only one explanation: the emission of greenhouse gases by human activity is causing profound changes to the climate system and to the world we live in.

The pace of change appears to be accelerating. Reports of sea levels rising faster than previously expected, of new temperature records, of an increasing toll of weather-related disasters, and anecdotal stories of impacts on livelihoods are accumulating. The year 2010 tied as the warmest year since records began in the 1850s, and threw up an astonishing series of extreme events.

Increases in global average air and ocean temperatures, widespread melting of snow and ice, and rising sea levels led the Intergovernmental Panel on Climate Change (IPCC) to report in 2007 that "warming of the climate system is unequivocal". As evidence continues to mount, that statement is even truer today.

USA: heat wave

In 2010, a very large area of the USA experienced high temperatures over an unusually long period. Downtown Los Angeles set an all-time record high temperature at 45°C in September 2010. Fires started in the hills and spread to residential areas.

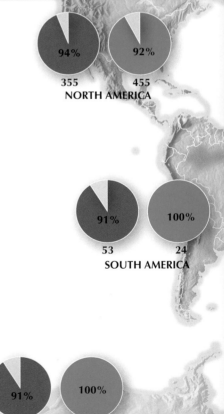

94%
355

92%
455

NORTH AMERICA

91%
53

100%
24

SOUTH AMERICA

91%
120

100%
24

ANTARCTICA

Russia: state of emergency

In 2010, a heat wave claimed 15,000 lives, with 7,000 deaths in Moscow alone. A state of emergency was declared in seven Russian regions, where tens of thousands of hectares of land was destroyed by fire, and hundreds of people were uprooted from their homes. As Russia's grain output was slashed by 40%, a grain export ban was imposed.

OBSERVED CLIMATE CHANGE IMPACTS

Number of significant observed changes and percentage of these changes that are consistent with climate change
1970–2004

- physical changes
- biological changes

China: floods

Flood waters from southwest to northeast China, including the municipality of Chongqing, shown here, led to the evacuation of 15 million people by the end of August 2010. Over 3,000 people died, and damage was estimated at over $50 billion.

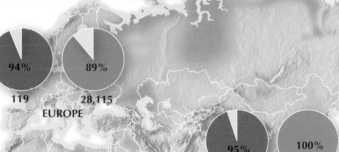

94%
119

89%
28,115
EUROPE

95%
106   **ASIA**   8

100%

100%
5   **AFRICA**   2

100%

100%
6
AUSTRALIA

94%
765

90%
28,671
WORLD

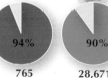

Australia: floods

December 2010 was the wettest on record for Queensland. The floods that resulted in January 2011 led to at least 22 deaths and affected more than 200,000 people. Taking into account the impact on the Australian economy, the cost is estimated in the region of $30 billion.

Pakistan: floods

Pakistan's monsoon rains in late July 2010 were unusually heavy, with 300 mm falling over the headwaters of the Indus in 24 hours. As the water moved downstream over the subsequent weeks, an estimated fifth of the country was inundated. More than 1,600 people were killed, and about 20 million displaced. Preliminary estimates placed the total damage at $15 billion.

2 POLAR CHANGES

Collapsing ice shelves
The collapse of the Larsen A Ice Shelf in 1995 was followed by that of the Larsen B Ice Shelf in 2002. Over the last century nearly 20,000 sq km (7,720 sq miles) of ice shelf was lost on the Antarctic Peninsula.

Warming in the Antarctic Peninsula and Arctic is driving large-scale melting of ice that will have both local and global consequences.

The presence of a hole in the ozone layer over the southern polar region has altered weather circulation patterns on Antarctica. It has brought more warm, moist, maritime air over the Antarctic Peninsula, contributing to warming and melting there, but has created a cooling effect in other areas. As the ozone hole recovers, that cooling effect is expected to diminish.

In East Antarctica, the changes are much less dramatic than those on the Peninsula, with some melting and thinning on coastal edges and some thickening in the interior. In West Antarctica, however, a coastal section of the ice sheet is now thinning quite rapidly.

Floating Arctic ice has covered the North Pole for millions of years. Its extent fluctuates with the seasons, but eight of the ten lowest extents have occurred in the last decade. The remaining ice is also thinner, with approximately 50 percent of the maximum recorded thickness having been lost by 2008. Already, the North Pole is free of ice in some summers.

In September 2007, the Arctic ice cap shrank to its smallest recorded extent, opening up the possibility of commercial shipping routes operating for the first time along the northern coasts of Canada and Russia. Some projections suggest that sea ice will disappear completely in the summer months by 2080.

While an open Arctic sea would facilitate shorter trade routes, industrial-scale fishing and the exploitation of minerals, it would be at great cost to the environment and to traditional livelihoods. A delay in the formation of the winter ice, an earlier break-up of ice in the spring, and thinner ice year round makes it hard for indigenous people using largely traditional methods to make a living.

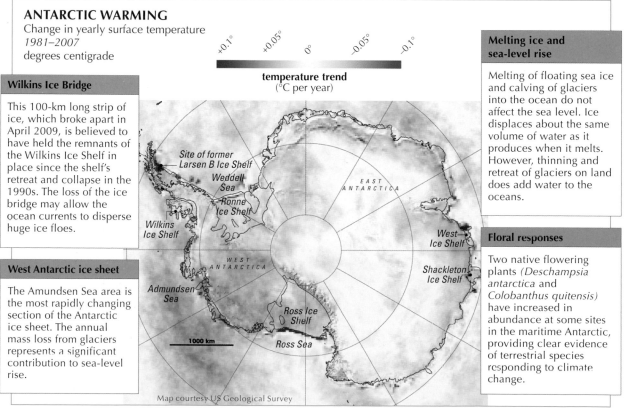

ANTARCTIC WARMING
Change in yearly surface temperature *1981–2007* degrees centigrade

+0.1° +0.05° 0° -0.05° -0.1°

temperature trend
(°C per year)

Wilkins Ice Bridge

This 100-km long strip of ice, which broke apart in April 2009, is believed to have held the remnants of the Wilkins Ice Shelf in place since the shelf's retreat and collapse in the 1990s. The loss of the ice bridge may allow the ocean currents to disperse huge ice floes.

West Antarctic ice sheet

The Amundsen Sea area is the most rapidly changing section of the Antarctic ice sheet. The annual mass loss from glaciers represents a significant contribution to sea-level rise.

Melting ice and sea-level rise

Melting of floating sea ice and calving of glaciers into the ocean do not affect the sea level. Ice displaces about the same volume of water as it produces when it melts. However, thinning and retreat of glaciers on land does add water to the oceans.

Floral responses

Two native flowering plants (*Deschampsia antarctica* and *Colobanthus quitensis*) have increased in abundance at some sites in the maritime Antarctic, providing clear evidence of terrestrial species responding to climate change.

Site of former Larsen B Ice Shelf
Weddell Sea
Ronne Ice Shelf
Wilkins Ice Shelf
EAST ANTARCTICA
West Ice Shelf
WEST ANTARCTICA
Shackleton Ice Shelf
Admundsen Sea
Ross Ice Shelf
Ross Sea
1000 km

Map courtesy US Geological Survey

ARCTIC

Minimum extent of Arctic summer sea ice

median for 1979–2000

September 2007

Greenland ice cap

area experiencing at least one melt day 2007

Potential shipping routes

northwest passage

northeast passage

What direction are we taking as an Inuit society? How is it we are going to deal with these monumental changes?

Sheila Watt-Cloutier, former Chair of Inuit Circumpolar Council

The permafrost around the Arctic is generally warming. In some areas, it is making a weakened coastline more prone to erosion, and causing subsidence, leading to the collapse of roads and buildings. It is also creating lakes of trapped melt water, which may increase carbon dioxide and methane emissions.

Each summer, parts of the Greenland ice sheet melt at the edges and on the surface. Although the melt area varies each year, the overall trend since 1979 has been upwards. Surface melt water finds its way through crevasses to the base of the ice, and forms a thin film between ice and bedrock. There are fears that this could increase the speed at which the ice sheet slides towards the sea.

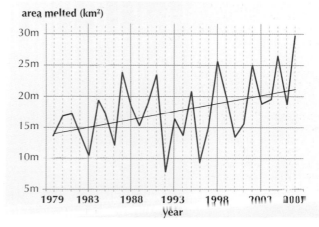

GREENLAND MELT
Rising trend in annual melt area
1979–2007

area melted (km²)

3 SHRINKING GLACIERS

THINNING
Regional average annual change in mass balance *1996–2005* meters of water equivalent

■ loss   ■ gain

0.65

0.90

0.26

0.72

0.49

0.25

0.43

0.65

1.14

Around the world, glaciers are losing mass and are in retreat.

The changes in glaciers over time provide valuable evidence of long-term climate change. The mass and extent of glaciers respond to temperature and snowfall in the very local geography of mountains and polar regions. Because the tell-tale signs of their expansion and retraction are clearly visible, scientists are able to draw conclusions about climate change from periods well before instrumental records became widespread.

Globally, glaciers have lost an average of more than half a meter (water equivalent) during the past decade or so. This is twice the rate of loss in the previous decade, and over four times the rate of loss in the late 1970s.

The front of most glaciers is receding to higher altitudes, and at such a rate that glaciologists, mountaineers, tourists and local residents are astonished by the changes that have occurred in the past decade. Tree stumps and even, in 1991, human remains that have been preserved in the ice for thousands of years, are now being revealed. Changing landscapes affect local plants and animals that colonize the newly exposed areas.

Glacial melting changes the flow of rivers, adding to water stress for millions of people. Lakes formed from melting glaciers are unstable, prone to abrupt collapse and flooding, threatening property and lives downstream.

THINNING OVER TIME
Average change in annual mass balance of global reference glaciers *1980–2008*

1980   1990   2000   2008

Alaska
Glaciers are both retreating and thinning.

Canadian Rockies
Tree stumps are being exposed for the first time in 2,500 years as glaciers recede.

Popocatepetl
The Ventorrillo glacier showed signs of retreat between 1950 and 1982.

USA
The South Cascades Glacier in Washington State has been retreating for a hundred years. From 1958 to 2005, its volume decreased by nearly half.

1979

2003

Northern Andes
The Quelccaya Glacier, Peru, is retreating 10 times more rapidly than it did in the 1970s and 1980s – by up to 60 meters a year.

Southern Andes
About half the glaciers surveyed in Chile show signs of retreat.

26   ◀◀ *24–25 Polar Changes*

GLACIAL RETREAT

Extent to which fronts of glaciers have moved
since 1950s
selected glaciers

■	almost all in retreat
■	more than half in retreat
■	some in retreat

Greenland

A rapid retreat and loss of ice mass in Greenland is giving cause for concern. Nearly 7 km² broke off on the night of 6 July 2010.

Scandinavia

Many glaciers are retreating, although increased snowfall is adding to their mass.

Himalayan and other Asian glaciers

Almost all glaciers surveyed are in retreat.

Tien Shan

The 400 glaciers in the north of the range have lost 25% of their volume since 1955. The glacier could be less than half its current volume by 2100.

European Alps

Glaciers have shrunk to a third of their 1850 extent and lost half of their volume. A typical example is the recession of the Morteratsch Glacier, Switzerland, between 1985 and 2007.

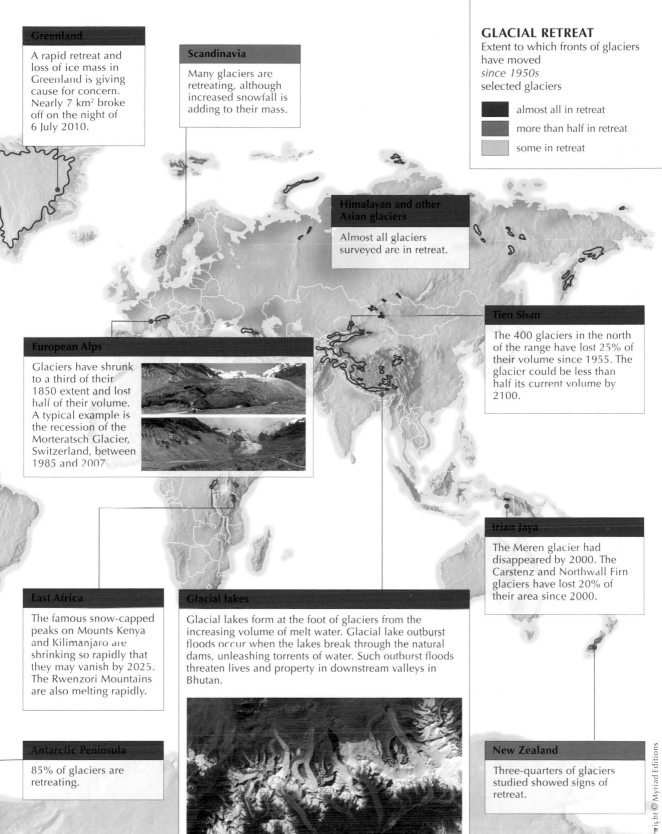

Irian Jaya

The Meren glacier had disappeared by 2000. The Carstenz and Northwall Firn glaciers have lost 20% of their area since 2000.

East Africa

The famous snow-capped peaks on Mounts Kenya and Kilimanjaro are shrinking so rapidly that they may vanish by 2025. The Rwenzori Mountains are also melting rapidly.

Glacial lakes

Glacial lakes form at the foot of glaciers from the increasing volume of melt water. Glacial lake outburst floods occur when the lakes break through the natural dams, unleashing torrents of water. Such outburst floods threaten lives and property in downstream valleys in Bhutan.

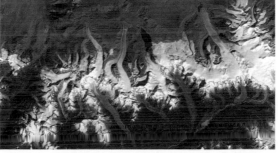

New Zealand

Three-quarters of glaciers studied showed signs of retreat.

Antarctic Peninsula

85% of glaciers are retreating.

4 OCEAN CHANGES

Coral reef bleaching
Coral reefs are the foundation for one of the world's most productive and diverse habitats so their health affects that of many other species. Heat stress is a major factor in coral reef bleaching – a process during which coral polyps expel the algae with which they have a symbiotic relationship, turn white, and may die.

Around the world, oceans are getting warmer and more acidic, affecting marine life.

Ocean temperatures, from the surface down to a depth of 700 meters, increased 0.1°C between 1961 and 2003. Temperature is fundamental to the basic life processes of organisms. It can influence metabolic rates and population growth of individual species and have broad repercussions on entire ecosystems. Coral reefs are particularly sensitive to temperature increases. Episodes of higher temperatures increase the frequency of coral bleaching and mortality.

Sea levels rose at an average rate of 1.8 ± 0.5 mm a year from 1961 to 2003 due to thermal expansion and melting of land-based ice. The recorded rate for the more recent period, 1993 to 2010, is much higher (3.3 ± 0.4 mm a year), generating concern among scientists that sea levels will rise faster than previously expected.

A separate issue is the contribution made by higher levels of carbon dioxide in the atmosphere to the acidity of ocean surfaces since the Industrial Revolution. Carbon dioxide in the atmosphere dissolves into the oceans and forms carbonic acid. In some sea areas there has been a 0.1 unit change in pH – corresponding to a 30 percent increase in acidity over levels in the mid-eighteenth century.

Increased acidity is expected to affect the variety of marine organisms with shells of calcium or aragonite, decrease oxygen metabolism of animals, and alter nutrient availability. The expected consequences of this change are already being observed in marine life. Scientists have measured decreases in the weight of the shells of small marine snails (pteropods) as well as decreases in the calcification of corals in the Great Barrier Reef. Impacts on these small organisms, which are the base of the food chain and therefore the foundation of productive habitats, could ripple up to affect fisheries and therefore protein and food security for millions of poor people.

An initial calculation of possible economic losses associated with a 10 to 25 percent decline in mollusk catches in the USA alone estimates losses for the year 2060 at between $324 million and $5.1 billion at current values. Under scenarios of increasing emissions of carbon dioxide, surface ocean pH is projected to decrease further, by 0.4 ± 0.1 pH units, becoming increasingly acidic by 2100 relative to pre-industrial conditions.

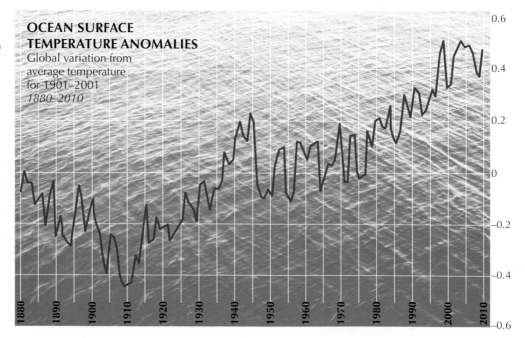

OCEAN SURFACE TEMPERATURE ANOMALIES
Global variation from average temperature for 1901–2001
1880–2010

OCEAN ACIDITY
Estimated mean sea-surface pH levels

8.45 8.40 8.35 8.30 8.25 8.20 8.15 8.10 8.05 8.00 7.95 7.90

Pre-industrial (pre-1750s)

1990s

CHANGING ACIDITY
Estimated change in annual mean sea-surface pH levels *between the pre-industrial period and the 1990s*

−0.05 −0.06 −0.07 −0.08 −0.09 −0.10 −0.11

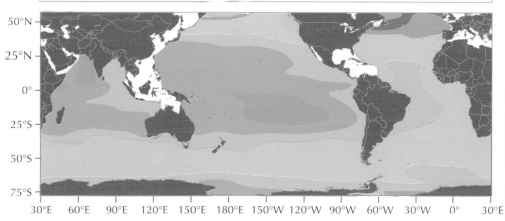

Maps courtesy of NOAA Pacific Marine Environmental Laboratory

IMPACT OF ACIDITY

Shell growth

1 micron

Incomplete shell due to acidity

Malformed shell growth due to acidity

Coccolithophorids are microscopic, one-celled organisms with calcareous shells vulnerable to increased acidity. They are the most abundant algal group in the world, and are important primary producers in marine foodwebs, serving as food for larger organisms such as snails and fish. As such, they are key parts of marine food webs, and their decline may have far-reaching consequences for ocean ecosystems.

46–47 Emissions Past & Present; 64–65 Food Security ▶▶   29

5 EVERYDAY EXTREMES

73% *of land area with adequate long-term records showed a significant increase in the number of warm night-time temperatures*

The frequency of some extreme events is increasing. A shift away from familiar patterns of climate variability is bringing changes in many aspects of climate.

In many parts of the world, the number of occasions on which precipitation is particularly heavy has increased. In China in 2010, heavy rains caused flooding and mudslides in 28 provinces. In Pakistan, unusually heavy rainfall over the watershed to the Indus River caused a large volume of water to move down the river over subsequent weeks, causing extensive flooding.

Since 1950, the number of heat waves has increased, affecting crop yields, human health, and the intensity of droughts. There has also been a widespread increase in the number of warm nights – a seemingly minor change, but one that reduces overnight relief from the day's heat.

It is very difficult to attribute one particular extreme, such as a single heavy rainfall or severe hurricane, to human-induced climate change rather than to the natural range of variability, but the increase in the probability of these events occurring can be linked to changes in the climate. In a sense, we can think of climate change as loading the dice in favor of these extreme events. That loading of the dice is easier to observe in large data sets, such as temperature and precipitation, which include decades of daily information from all over the world. It is much more difficult to detect change in trends of comparatively rare events. For instance, tropical cyclones and hurricanes wreak havoc around the world, yet there is only low confidence at present that their frequency and intensity have increased since the 1970s.

As examples of the impacts of recent extreme weather events, the still-recovering city of New Orleans or the rural population suffering the long-term effects of the Pakistan floods, can provide insight into the types of losses that might be increasingly experienced as the planet warms.

But these impacts cannot all be blamed on climate change. Many other social and economic factors contribute to hazard vulnerability and loss. Poverty, poor warning systems and land-use planning, inadequate shelter, and other issues of economics, education, household resources, and governance contribute to vulnerability to these extremes.

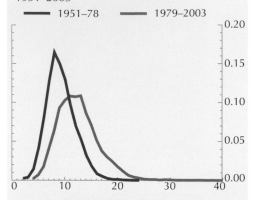

INCREASE IN WARM NIGHTS
Based on temperature records from 895 grids
1951–2003

— 1951–78 — 1979–2003

This probability density function compares the number of warm nights observed in two different periods, using gridded data for areas with adequate records. A warm night is defined as one with minimum temperatures in the top 10th percentile of temperatures for that location. In more recent decades, the likelihood of a year with more warm nights increased.

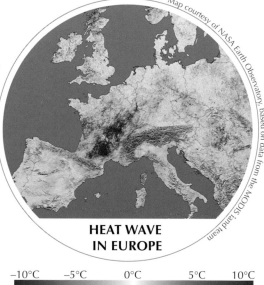

Map courtesy of NASA Earth Observatory, based on data from the MODIS land team

HEAT WAVE IN EUROPE

−10°C   −5°C   0°C   5°C   10°C

Difference between land surface temperatures *in July 2003 and July average for 2000–02*

Parts of Europe experienced a major heat wave in late July and early August 2003, which resulted in approximately 40,000 deaths.

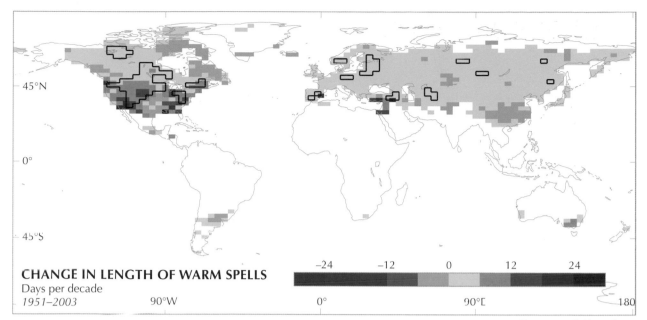

CHANGE IN LENGTH OF WARM SPELLS
Days per decade
1951–2003

| -24 | -12 | 0 | 12 | 24 |

90°W   0°   90°E   180

Climate models project that there will be more heat waves in the future. The above analysis of temperatures from 1951 to 2003 provides observational support for models. Here, heat waves, are termed "warm spells", which are defined as the number of days per year with at least six consecutive days of maximum temperatures in the top 10% of all maximum temperatures for that time of year.

Areas in the yellow to red shades indicate a trend to longer warm spells. Black lines enclose areas where statistical analysis indicates that the trend is significant at the 5% level.

Trends were not calculated for areas with fewer than 40 years of data, and data that ended before 1999, so we do not have information for much of Africa, South America, the Middle East, Asia, and Australia.

Map courtesy L. Alexander, Climate Change Research Centre, University of New South Wales

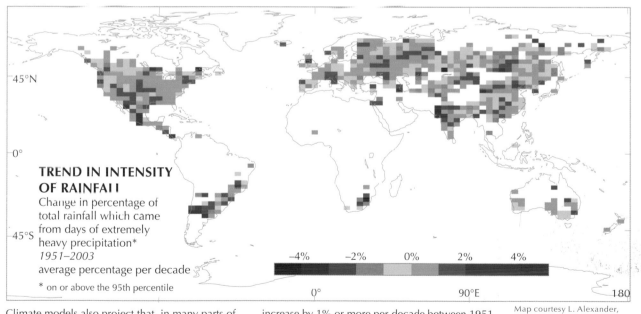

TREND IN INTENSITY OF RAINFALL
Change in percentage of total rainfall which came from days of extremely heavy precipitation*
1951–2003
average percentage per decade

* on or above the 95th percentile

| -4% | -2% | 0% | 2% | 4% |

0°   90°E   180

Climate models also project that, in many parts of the world, rain will tend to fall in more intense bouts. This map represents analysis indicating that in much of the USA, and Europe, as well as parts of Asia, the amount of total annual rainfall contributed by the heaviest 5% of rainy days has tended to increase by 1% or more per decade between 1951 and 2003. Areas in white show lack of adequate data sets for analysis.

There is less area with statistically significant trends in intense rainfall than in longer warm spells.

Map courtesy L. Alexander, Climate Change Research Centre, University of New South Wales

PART 2 The Changing Climate

The Earth's climate system is a set of complex interactions between land, water, and air, reflecting the interplay of social, economic, and institutional factors. And these interactions operate over a huge span of spatial and temporal scales. Our understanding of climate change will always be incomplete.

If we take the image of a jigsaw puzzle, the evidence for climate change presented in this atlas fills in three sides. The first, perhaps the foundation of our picture at the bottom of the puzzle, is the physical understanding of the effect of increasing greenhouse gas concentrations on radiation and the Earth's climate system, well documented for over a century. Imagine the atmosphere as a thermal blanket, such as a comforter or duvet that keeps our sleeping bodies warm by absorbing the heat radiating from us. If we fill the holes in the duvet with extra feathers, it will keep us even warmer. This is what we are doing to the Earth's atmosphere – adding greenhouse gases and keeping more heat in the atmosphere.

The left boundary of the puzzle is the evidence of paleo-climate and historic climate change, as revealed in enormous volumes of data. Observations from around the world continue to record changes consistent with scientists' expectations.

The right edge of the puzzle is the computer modeling, which brings together the physical understanding and observations. All the models show global warming in response to increasing greenhouse gas concentrations. Natural factors alone cannot explain the observed global warming over the last few decades or so; in order to match the observations, it is necessary to include the role of greenhouse gases in trapping increased radiation.

The top edge of the jigsaw puzzle is incomplete. In this third edition, we report on the potential for large-scale, possibly catastrophic, changes in climate. The very nature of climate change – in what we call a coupled socio-ecological system – is the potential for surprise. Crises in our capacity to respond to climate change and in potential shifts to quite different climatic conditions cannot be ruled out. The upper end of potential climate change is still poorly charted. Chaos theory in complex systems is a fascinating area: small changes lead to diverging outcomes, making long-term predictions impossible.

Confidence in this outline of climate change has not come easily. With new monitoring systems, state-of-the-art computers, and thousands of experts around the world, we continue to observe, analyze, debate, and refine areas of understanding, albeit with much more media coverage. For the past 15 years, the Intergovernmental Panel on Climate Change has concluded, with increasing confidence, that we are changing the global climate. It is no longer "do we believe" in computer projections; the evidence is in front of our daily lives. Increasingly, the evidence points to a commitment to at least 2°C of global climate change. That is raising alarm bells.

Despite the large uncertainties in many parts of the climate science and policy assessments to date, uncertainty is no longer a responsible justification for delay in either adaptation or mitigation policies.

Dr Stephen H Schneider
(1945–2010)
Melvin and Joan Lane Professor for Interdisciplinary Environmental Studies, Stanford University
May 20, 2010

33

6 THE GREENHOUSE EFFECT

The intensification of the greenhouse effect is driving increases in temperature, and many other changes in the Earth's climate.

Without the natural atmospheric greenhouse effect, which captures and holds some of the sun's heat, humans and most other life-forms would not have evolved on Earth. The average temperature would be –18°C, rather than 15°C.

The greenhouse effect operates in the following way. Solar radiation passes through the atmosphere, and heats the surface of the Earth. Some of that energy returns to the atmosphere as long-wave radiation or heat energy. Another portion of that energy is captured by the layer of gases that surrounds the Earth like the glass of a greenhouse, while the rest passes into space. Changes to the composition of this layer of gases are central to climate change.

Over the last 250 years or so, human activity – such as extensive burning of fossil fuels, the release of industrial chemicals, the removal of forests that would otherwise absorb carbon dioxide, and their replacement with intensive livestock ranching – has changed the types and amounts of gases in the atmosphere, and substantially increased the capacity of the atmosphere to absorb heat energy and emit it back to Earth. The major greenhouse gases augmented by human activities are carbon dioxide, tropospheric ozone, nitrous oxide, and methane. Other industrial chemicals, including many halocarbons, also add to the effect.

Some of these gases only stay in the atmosphere for a few hours or days, but others remain for decades, centuries, or millennia. Greenhouse gases emitted today will drive climate change long into the future, and the process cannot be quickly reversed. In addition, warming may cause changes, or "feedbacks", that further accelerate the greenhouse effect. For instance, if the highly reflective snow cover decreases, more solar radiation will warm the Earth's surface. The warmer the Earth, the more heat energy is emitted back to the atmosphere. And if warming leads to extensive thawing of permafrost, there may be a large release of methane, a potent greenhouse gas.

THE GREENHOUSE EFFECT

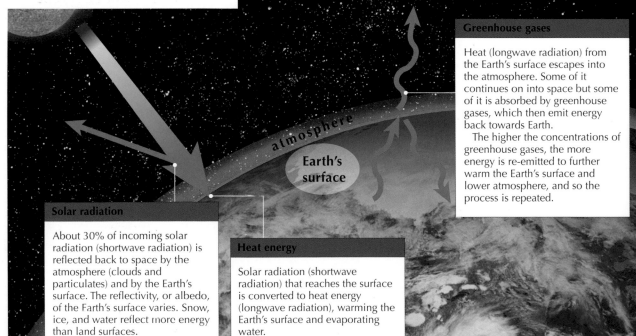

Greenhouse gases

Heat (longwave radiation) from the Earth's surface escapes into the atmosphere. Some of it continues on into space but some of it is absorbed by greenhouse gases, which then emit energy back towards Earth.

The higher the concentrations of greenhouse gases, the more energy is re-emitted to further warm the Earth's surface and lower atmosphere, and so the process is repeated.

atmosphere

Earth's surface

Solar radiation

About 30% of incoming solar radiation (shortwave radiation) is reflected back to space by the atmosphere (clouds and particulates) and by the Earth's surface. The reflectivity, or albedo, of the Earth's surface varies. Snow, ice, and water reflect more energy than land surfaces.

Heat energy

Solar radiation (shortwave radiation) that reaches the surface is converted to heat energy (longwave radiation), warming the Earth's surface and evaporating water.

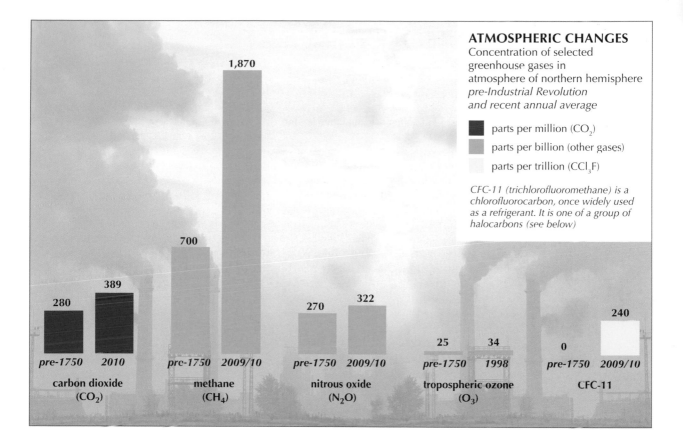

ATMOSPHERIC CHANGES

Concentration of selected greenhouse gases in atmosphere of northern hemisphere *pre-Industrial Revolution and recent annual average*

- ■ parts per million (CO_2)
- ■ parts per billion (other gases)
- □ parts per trillion (CCl_3F)

CFC-11 (trichlorofluoromethane) is a chlorofluorocarbon, once widely used as a refrigerant. It is one of a group of halocarbons (see below)

	pre-1750	recent
carbon dioxide (CO_2)	280	389
	pre-1750	*2010*
methane (CH_4)	700	1,870
	pre-1750	*2009/10*
nitrous oxide (N_2O)	270	322
	pre-1750	*2009/10*
tropospheric ozone (O_3)	25	34
	pre-1750	*1998*
CFC-11	0	240
	pre-1750	*2009/10*

APPROXIMATE CONTRIBUTION OF GHGS

to positive radiative forcing of the atmosphere *post-1750*

- nitrous oxide (N_2O) 5%
- halocarbons 11%
- tropospheric ozone (O_3) 12%
- methane (CH_4) 16%
- carbon dioxide (CO_2) 56%

Halocarbons, are a group of climate-forcing gases that includes CFCs, and their replacement compounds HCFCs and HFCs.

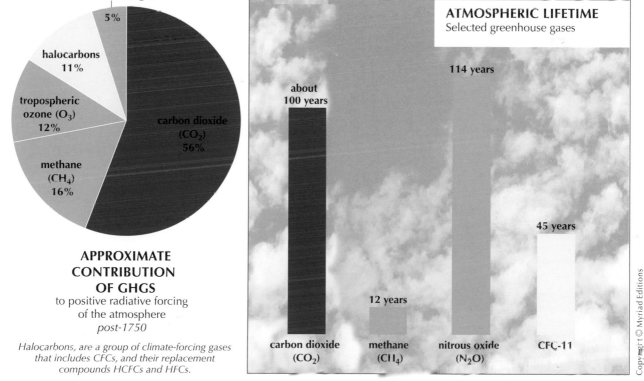

ATMOSPHERIC LIFETIME

Selected greenhouse gases

- carbon dioxide (CO_2) — about 100 years
- methane (CH_4) — 12 years
- nitrous oxide (N_2O) — 114 years
- CFC-11 — 45 years

7 THE CLIMATE SYSTEM

The climate system is the highly complex system consisting of the atmosphere, the hydrosphere, the cryosphere, the land surface and the biosphere, and the interactions between them. The climate system evolves in time under the influence of its own internal dynamics and because of external forcings.

IPCC

The climate system is adjusting to an increase in the heat trapped in the Earth's atmosphere caused by greenhouse gases.

The Earth's climate system functions as a giant "heat distribution engine." Atmospheric and oceanic circulations move heat energy around the world, in an effort to distribute it more equally. This creates the long-term average conditions referred to as "climate", within which we experience the short-term, day-to-day variability of "weather." The addition of more heat energy to the atmosphere and the Earth's surface is expected to alter the climate system's heat distribution patterns, to change many average climate conditions and thus create more variation in the weather.

Most solar radiation and surface heating occurs at the equator, where the sun's rays are nearly perpendicular to the surface all year round. The poles receive much less radiation because of the Earth's orbit and tilt relative to the sun. Atmospheric and oceanic circulations contribute equally in moving energy from the equator towards the poles. The large-scale weather systems that generate the migrating warm and cold fronts, and their associated storms, are part of this process. The climate is also influenced by processes that contribute to multidecadal variability, such as El Niño and others in the Atlantic and Pacific.

The process of heat energy distribution by the global climate system is largely responsible for regional climates, and an increase in temperature differentials between the tropics and the poles could disrupt the climate in many ways. Warmer summers, heat waves, drier winters, less snowfall, and changing frequency and intensity of storms, are all possible results.

ATMOSPHERIC CIRCULATION

At the equator, solar radiation heats the land and the oceans, evaporating water. Warm air rises, carrying water vapor and creating heavy rainfall in the atmosphere. The resulting drier, colder air is pushed away from the equator and descends to Earth in the Tropics of Cancer and Capricorn, where deserts form. Trade winds blow back to the equator to complete the circulation. Similar circulation cells exist over the mid-latitudes and polar regions.

Jet stream

The typical flow of the mid-latitude jet stream or polar front may be altered by climate change, affecting temperature and precipitation patterns.

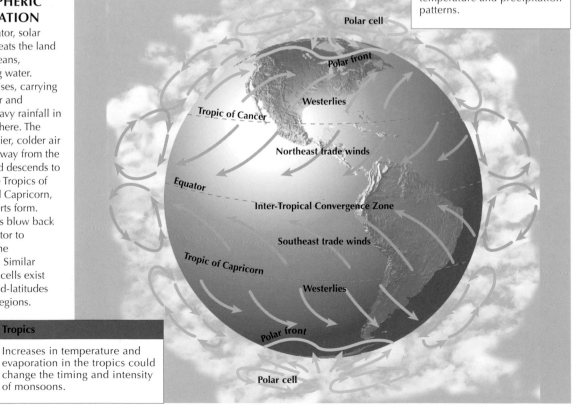

Polar cell

Polar front

Westerlies

Tropic of Cancer

Northeast trade winds

Equator

Inter-Tropical Convergence Zone

Southeast trade winds

Tropic of Capricorn

Westerlies

Polar front

Polar cell

Tropics

Increases in temperature and evaporation in the tropics could change the timing and intensity of monsoons.

OCEAN CIRCULATION

The ocean currents distribute energy from solar heating. Some currents are mainly driven by winds and tides, but others are primarily driven by differences in ocean temperature and salt concentration that affect the density of seawater. Water naturally mixes to even out the distribution of heat and salt. Circulation occurs as the denser, colder, saltier water drops below the warmer, fresher water. This drives the "thermohaline circulation", or the "ocean conveyor belt", although that metaphor doesn't fully capture the complexity of mixing processes.

If climate change warms the polar waters and/or decreases their salinity by adding fresher water from melting glaciers, the difference in water density will decline and the circulation pattern is expected to slow down or even collapse. Based on current models, many scientists view the potential collapse as a "low-probability, high-impact" scenario in the 21st century, although some see greater probabilities and potential in the 22nd century.

Even a slowing of the circulation is likely to have a wide-ranging impact. Descending cold waters carry carbon dioxide into deep water, away from the atmosphere. Elsewhere, parts of the world rely on upwelling waters to bring nutrients from the bottom of the ocean to the surface, where they help support the local ecology and fisheries. Temperature and rainfall patterns are also expected to change with alterations to this circulation pattern.

Storms

Increased temperatures and water vapor in the air over tropical oceans will create improved spawning conditions for cyclones, hurricanes and typhoons, although other climate circulation features are required for formation. Modeling studies suggest storms may have greater wind speeds and precipitation.

Northern Europe

Europe has a mean annual surface air temperature some 5°–7°C warmer than regions at the same latitude in the Pacific. A collapse, or even a slowing, of the Gulf Stream would reduce the amount of warm, tropical water brought north. Although this cooling might be offset by atmospheric warming, the consequences could be severe.

El Niño

The frequency and intensity of El Niño events, triggered when warm water in the Pacific extends east, may be affected by changes in climate systems. During El Niño years, rainfall follows the warm water, typically leading to flooding in Peru, drought in Indonesia and Australia, and a disruption to climate patterns elsewhere in the world.

Europe

Recent heat waves in Europe may be linked to the global ocean circulation.

Sahel

Drought, changes in seasonal rainfall and flooding have been linked to changes in ocean currents and other multi-decadal patterns of climate variability.

Sea-to-air heat transfer

Atlantic Ocean

Gulf Stream

Indian Ocean

Solar warming of oceans

Pacific Ocean

Gulf Stream

warm shallow current

cold and salty deep current

8 INTERPRETING PAST CLIMATES

ACCUMULATED KNOWLEDGE

Total number of scientific articles referring to climate change and published in journals indexed in Web of Science
1970–2010

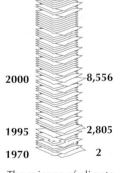

2010 — 47,358

2005 — 19,380

2000 — 8,556

1995 — 2,805

1970 — 2

The science of climate change is well established, with over 5,000 articles now appearing every year in peer-reviewed journals. Over 3,000 scientists in the Intergovernmental Panel on Climate Change (IPCC) collect and review this knowledge.

Concentrations of carbon dioxide and methane are higher than they have ever been in the last 800,000 years. The Earth is warmer than at any time in the past 1,000 years.

Records of the Earth's past climates, reconstructed from ice cores, tree rings, paleoclimatic fingerprints in ocean and lake sediments, cave deposits, glaciers, and even ship records of sea-surface temperatures, confirm that global warming is real and unprecedented. The Earth is warmer than in the past millennium, and the commitment to future warming is evident in the record levels of carbon dioxide (CO_2) in the atmosphere. Modern human societies and economies have never faced such conditions.

A physical explanation of the greenhouse effect was already well developed 100 years ago, and since then scientists have identified trends in atmospheric composition and temperature extending over hundreds of thousands of years. The concentrations of carbon dioxide and methane in the atmosphere are now at the highest levels for over 800,000 years. The

atmospheric concentration of carbon dioxide has risen from 315 parts per million (ppm) in the 1950s to just over 388 ppm in 2010. Paleoclimate records indicate that the pattern of global temperature fluctuations has been similar to the pattern of change in CO_2 levels.

The physical understanding of the climate system, with its many variables, is captured in computer models. Their efficacy is tested against the record of past climates, and many have managed to reconstruct past climates reasonably accurately. Fluctuations in past temperatures have been shown to be caused by natural forces, such as cycles of solar energy, changes in the Earth's orbit, and volcanic eruptions that send gases and dust into the atmosphere. However, the variability and trends in historical global temperatures can only be explained if both natural forces and greenhouse gas emissions from human activity are included in the models. This validation of models of the physics of the climate system gives scientists the confidence to use these models to project future climate change.

CO_2 FLUCTUATIONS
400,000 years ago to present day

— concentration in ice-core samples
— concentration in atmosphere

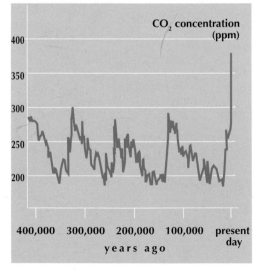

LINK BETWEEN CO_2 AND TEMPERATURE
160,000 years ago to present day

— CO_2 concentration in atmosphere
— global temperature

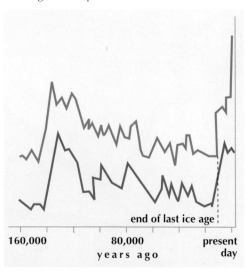

SURFACE TEMPERATURES
Information from a range of
measurements
900–2000

—— combined proxy data set 1
—— combined proxy data set 2
—— tree rings
—— combined proxy data set 3
—— borehole temperatures
—— glacier lengths

increasing certainty

Efforts to reconstruct the history of past surface temperatures for the Earth have generated considerable controversy. The long period of less variability followed by an upturn in the last century, shown in a figure known as the "hockey stick curve", was taken as definitive evidence of human-induced climate change. In response to the controversy, the US Academy of Sciences was asked to assess the state of science in this area. A key figure summarizing part of the assessment illustrates their conclusion that multiple sets of proxy information provide a qualitatively consistent view of temperature changes over the past 1,100 years. The last 400 years show particularly strong agreement, as reflected in the background shading of the diagram, which indicates increasing levels of certainty.

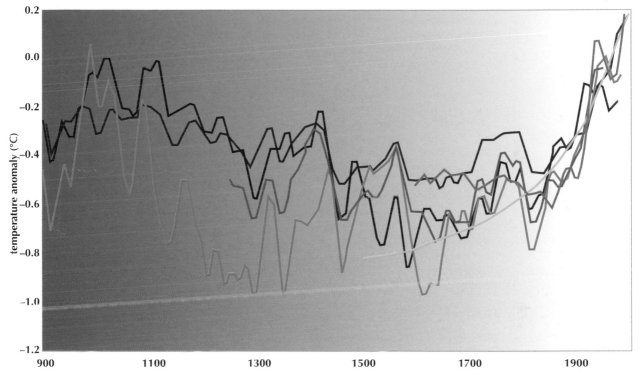

ACCOUNTING FOR WARMING
Comparison of observed changes in
surface temperature with results
simulated by climate models
using natural and
anthropogenic forcings
1906–2005

—— decadal averages
of observations

5%–95% range
for climate models using
only natural forcings

5%–95% range
for climate models using
both natural and
anthropogenic
forcings

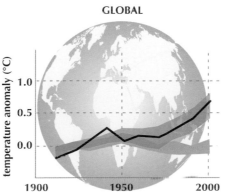

GLOBAL

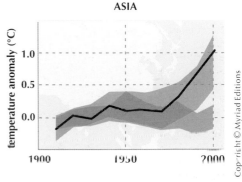

ASIA

Copyright © Myriad Editions

SCIENTIFIC
CONFIDENCE
IN PREDICTIONS
OF CLIMATE
CHANGE

SCIENTIFIC CONFIDENCE IN PREDICTIONS OF CLIMATE CHANGE
2007

Virtually certain

- Higher temperatures: warmer days, more frequent hot days and nights, fewer cold days and nights, over most land areas.

Very likely

- More frequent warm spells and heat waves, over most land areas.
- More intense rain, over most areas.

Likely

- Increased area affected by drought.
- More frequent intense tropical cyclones: risk of stronger winds and heavier rain.
- More frequent coastal flooding due to extreme high sea levels.

Climatologists are confident about many aspects of future climate change based on: the physical understanding of the climate system, trends in observed climates, and projections from computer models of the climate.

Projections of future climates start with scenarios that have made different assessments of fossil fuel use, the development and distribution of technologies that reduce emissions, and the speed of population and economic growth. These lead to different pathways of economic and social development, and therefore to different predictions of future energy use.

Concerns about the best policy to combat climate change have led to an additional suite of emissions scenarios based on reaching a target level of climate change. This requires working backwards from a maximum global temperature limit, say 2°C, to the level of greenhouse gas concentrations that could be tolerated, and then to the limit that would need to be placed on emissions in order not to exceed that concentration. There is considerable uncertainty in each of these steps.

The pre-industrial concentration of carbon dioxide (CO_2) was 280 parts per million in the atmosphere. Most emissions scenarios expect a concentration of over 520 parts per million by 2100 in the absence of concerted climate policy. Policy targets in many countries are for reductions of 50 to 80 percent in emissions by 2050.

By the 2050s, the global average temperature is almost certain to be more than 1°C higher than that for 1960 to 1990, and may be as much as 6°C higher by the end of this century. Warming in the higher latitudes and polar regions is likely to be greater than the global averages. Changes in precipitation patterns are, however, less certain.

Forecasts at the local level, including extreme events such as floods and cyclones, are less certain. Modeling the extraordinary complexity of the global climate system has to take into account shifts in regional systems, such as mid-latitude high pressures, and local topography, such as the rain shadow in the lee of mountain ranges. However, climate scientists are beginning to explore the range of plausible shifts in regional variables such as sunshine, humidity, precipitation, and wind.

EMISSIONS SCENARIOS
Projected annual emissions
1990–2100

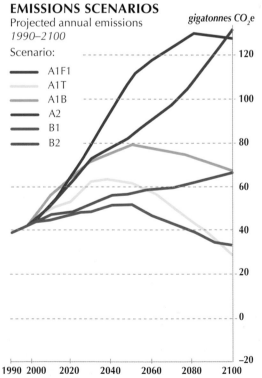

STABILIZATION SCENARIOS
Levels to which global emissions would need to be constrained in order to arrive at different concentrations of CO_2e in atmosphere

- 855 – 1,130 ppm CO_2e
- 710 – 855
- 590 – 710
- 535 – 590
- 490 – 535
- 445 – 490

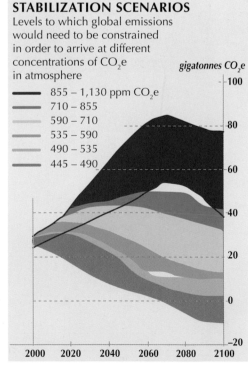

LOCAL WARMING
Annual temperature change resulting from a doubling of CO_2 equivalents in atmosphere

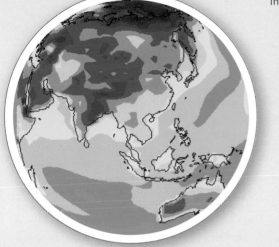

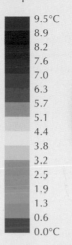

9.5°C
8.9
8.2
7.6
7.0
6.3
5.7
5.1
4.4
3.8
3.2
2.5
1.9
1.3
0.6
0.0°C

Globes courtesy of David Stainforth, www.ClimatePrediction.net

Future climate scenarios are compiled by running a computer model several times, each with slightly different assumptions. Each run can be compared with the climate of the past 100 years as a way of assessing its potential relevance for modeling the future. Runs are done on supercomputers, or using networks of personal computers around the world.

Almost all land areas are likely to get warmer. Globally, precipitation is likely to increase, but the pattern of where it will get wetter or drier is uncertain. For example, in most models, June to August gets wetter in East Africa and East Asia and drier in the Mediterranean, while in central North America some models produce wetter and some drier summers (see below).

PRECIPITATION
Model simulations of wetter or drier June to August with a doubling of CO_2 equivalents

— Mediterranean Basin
— Central North America
— East Africa
— East Asia

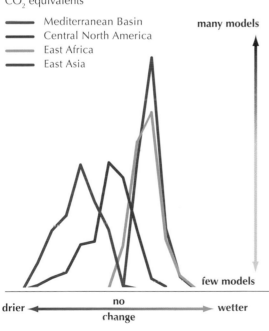

many models

few models

drier ← → wetter

no change

PROJECTED WARMING
Increase in average global temperatures
2000–2100
degrees Celsius

■ average of global climate models

■ area of uncertainty

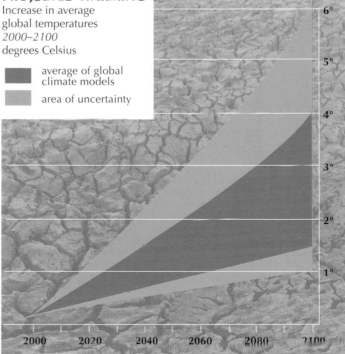

6°
5°
4°
3°
2°
1°

2000   2020   2040   2060   2080   2100

10 CLIMATE & SOCIAL CRISES

The global burden of migration related to climate change might be

100 million

people by 2050

Climate change results from complex interactions in the natural environment, coupled with social and economic changes. Such complexity is not fully understood, and is impossible to predict.

As a result of climate change, dramatic alterations – often referred to as tipping elements – could occur in the Earth's biophysical system. Equally, countries unable to adapt to the impact of climate change could experience social and political upheaval. The combination of sudden changes in climate conditions, impacts, and resources, and in social, economic, and institutional conditions has the potential to create large-scale humanitarian crises in the future.

Tipping elements force the Earth's climate to switch from its present condition to a very different state. Some tipping elements can be rather rapid; others are likely to take a century or longer to complete. One of the most worrisome is the release of frozen methane from deep-sea deposits which, combined with the melting of permafrost, would create sufficient methane to spike global temperatures by tens of degrees.

Many countries are already fragile; their economies are sensitive to the impacts of droughts and floods. Even some political changes can be linked to climatic stresses. The sharp increase in food prices in late 2010 was one trigger to the Arab Spring in North Africa the following year.

Preventing "dangerous" climate change is a central tenet of the United Nations Framework Convention on Climate Change (UNFCCC). Since 2006, there has been increasing concern that global warming of 2°C represents a major threshold in risk. International negotiations have set 2°C as a common target for maximum global climate change, as indicated in the Copenhagen Accord. However, African countries note that even this level is dangerous in Africa, and propose a global target of 1.5°C.

The science is very uncertain, and the road ahead likely to be full of surprises – a feature of complex, coupled, socio-ecological systems such as the Earth's climate. However, although these are complex issues that will take years to understand adequately, the analysis of potential biophysical, social and economic crises is growing rapidly. In the case of planetary crises, we know there are potential threats of grave consequence.

TIPPING ELEMENTS
Some examples

Global warming threshold	Tipping element	Time period over which change might occur
3° – 6°	El Niño-Southern Oscillation enters a near-permanent phase	100 years
3° – 5°	Sahara and West African monsoon: abrupt shift in climate pattern	10 years
3° – 5°	Boreal forests: exposed to fires and pests	50 years
3° – 5°	Atlantic Ocean: melting ice shifts Gulf Stream southward	100 years
3° – 5°	West Antarctic Ice Sheet: collapses	300 years
3° – 4°	Amazon rainforest: dieback due to reduction in rainfall	50 years
1° – 2°	Greenland Ice Sheet: accelerated melting leading to ice-sheet break up	300 years
0.5° – 2°	Ice-free Arctic in summer absorbs more heat and accelerates warming	10 years

Relocating vulnerable communities

Migration is the only realistic adaptation for many coastal communities in the Arctic region, to avoid melting permafrost and coastal erosion. The communities themselves – some call them the first climate change migrants – wish to move, but insist that they maintain social, cultural, and economic cohesion.

FRAGILE STATES AND CLIMATE CRISES
Countries with a high risk of political instability where climate change could lead to humanitarian crises
2007

- high risk of armed conflict
- high risk of political instability
- no risk

Drought and complex emergencies

There is a long history of migration and conflict in Darfur, western Sudan, related to the surges of drought but also to the politics and economy of the region. While climate change may increase such stresses, it is difficult to prove that they are a consequence of it. Whatever the immediate cause, these are desperate humanitarian crises for which we have few lasting solutions.

Coastal insecurity

The coastal zones of Bangladesh and India are vulnerable to erosion and storm surges. People are moving away each time a piece of the island is flooded or disappears into the sea. Many move to urban areas, such as Dhaka and Calcutta, adding to their environmental burdens. Ultimately it is a human rights issue.

PART 3 Driving Climate Change

The climate system is relentlessly changing. The immediate causes of climate change are the emissions of greenhouse gases released from energy production and consumption, agriculture, transport and ecological processes. Behind these sources of greenhouse gas emissions are broader driving forces related to economic transformations, and to prospects for alternative energy pathways and equity across regions and populations.

Industrialized nations are responsible for most of the past emissions, and remain the major sources today. Focusing on energy needs for economic development, rather than for sustainability, industrialized countries invested heavily in carbon-intensive technologies, such as coal-fired power plants, massive road systems, and electrical grids. Globalization has made extensive transportation of goods and services the norm, and access to foreign markets crucial for economic growth.

However, by 2005 China had passed the USA as the world's largest emitter of greenhouse gases. And the top seven emitters – China, USA, the European Union, Brazil, Indonesia, Russia, and India – account for nearly 60 percent of emissions. Of course, industrialized countries still head the league tables on a per capita basis.

The legacy of greenhouse gases – some remain in the atmosphere for centuries – guarantees the inevitability of climate change for decades to come. Developing countries, even though they have contributed little to historical emissions, will suffer an enormous burden of the impacts.

This imbalance between responsibility for the current causes of climate change and its impacts creates an enduring global inequity. Booming economies in countries such as China and India lead to greater energy use and higher living standards; least develop countries struggle to afford national electricity grids, much less energy-intensive luxuries.

The pace of change may be accelerating. Despite the economic slowdown, emissions continue to rise, possibly on a trajectory that a few years ago was considered at the high end of plausible scenarios.

The entrenched commitment to carbon-intensive economies, the relationship between luxury consumption and basic needs, and realignment in the global political economy are all part of the challenges in achieving reductions in emissions.

For many, emissions of greenhouse gases are necessary for survival; for others, they result from luxury consumption and lifestyles.

This is not fiction, this is science. Unchecked, climate change will pose unacceptable risks to our security, our economies, and our planet.

Barack Obama
December 18, 2009
at COP15, Copenhagen

11 EMISSIONS PAST & PRESENT

SHARE OF ANNUAL EMISSIONS
From energy use
2006

**Total
28,436 million
tonnes CO$_2$e**

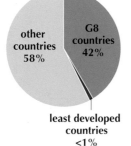

other countries
58%

G8 countries
42%

least developed countries
<1%

Most greenhouse gases have been, and continue to be, emitted to meet the needs of modern industrial societies.

Once in the atmosphere, most greenhouse gases remain for dozens of years. Carbon dioxide, the major contributor, remains for about 100 years. Over half of all current greenhouse gas emissions are from the energy used in heating and lighting, transportation and manufacturing. Countries with a long history of industrialization have contributed the majority of the greenhouse gases in the atmosphere. These, and emissions made from this point forward, will influence the future of the climate for many years to come.

In the future, countries currently undergoing industrialization will account for a higher percentage of annual greenhouse gas emissions than previously but it will be many years before their accumulated emissions match those of today's most industrialized nations.

This difference in past and future contributions to the overall levels of greenhouse gases raises important equity issues that are at the heart of international negotiations over how best to mitigate and adapt to climate change.

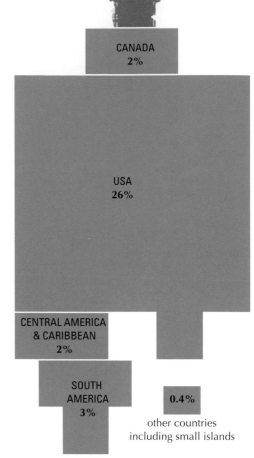

CANADA
2%

USA
26%

CENTRAL AMERICA & CARIBBEAN
2%

SOUTH AMERICA
3%

0.4%

other countries
including small islands

CO$_2$ IN THE ATMOSPHERE
Global tropospheric concentration
1750–2010
parts per million

ice-core measurements ▬ atmospheric readings

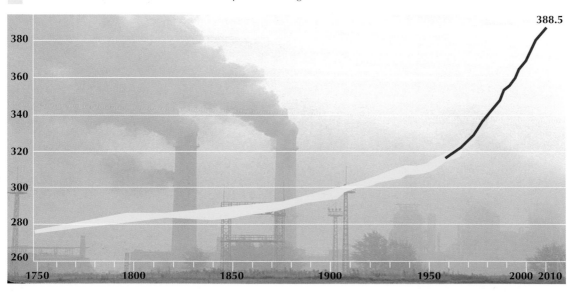

388.5

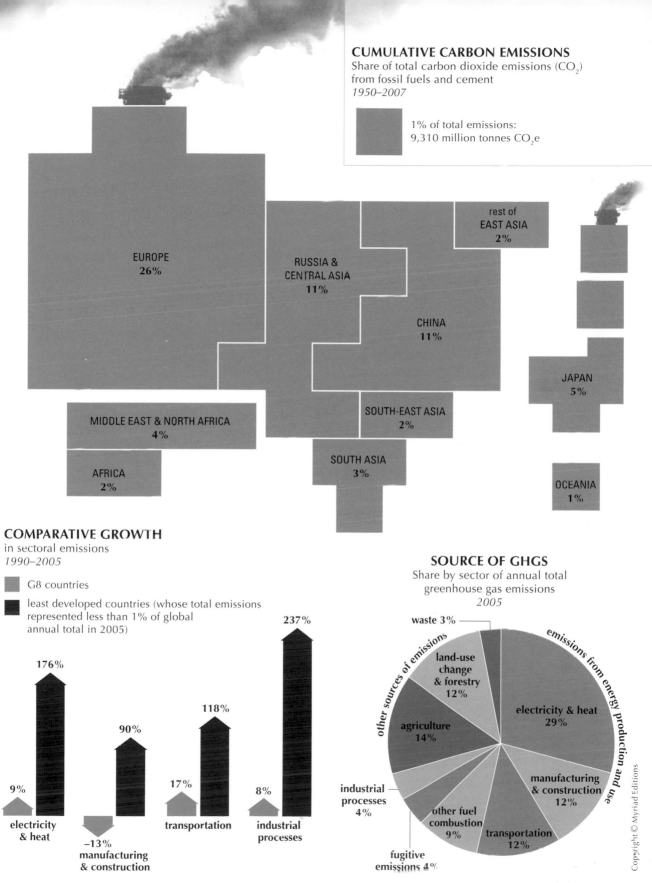

CUMULATIVE CARBON EMISSIONS
Share of total carbon dioxide emissions (CO_2) from fossil fuels and cement
1950–2007

1% of total emissions:
9,310 million tonnes CO_2e

EUROPE
26%

rest of
EAST ASIA
2%

RUSSIA &
CENTRAL ASIA
11%

CHINA
11%

JAPAN
5%

MIDDLE EAST & NORTH AFRICA
4%

SOUTH-EAST ASIA
2%

AFRICA
2%

SOUTH ASIA
3%

OCEANIA
1%

COMPARATIVE GROWTH
in sectoral emissions
1990–2005

G8 countries

least developed countries (whose total emissions
represented less than 1% of global
annual total in 2005)

9% 176%
electricity
& heat

–13% 90%
manufacturing
& construction

17% 118%
transportation

8% 237%
industrial
processes

SOURCE OF GHGS
Share by sector of annual total
greenhouse gas emissions
2005

waste 3%

other sources of emissions

land-use
change
& forestry
12%

agriculture
14%

emissions from energy production and use

electricity & heat
29%

manufacturing
& construction
12%

industrial
processes
4%

other fuel
combustion
9%

transportation
12%

fugitive
emissions 4%

CARBON DIOXIDE EMISSIONS GROWTH

Average annual rate of increase
1990s & 2000–08

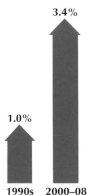

3.4%

1.0%

1990s   2000–08

The emission of greenhouse gases from the burning of fossil fuels continues to be the major cause of climate change.

Over three-quarters of carbon dioxide (CO_2) emissions, derive from the burning of fossil fuels such as oil, natural gas, and coal, primarily for electricity generation, transportation, heating, and cooling. Further emissions of CO_2, and of other GHGs, arise from industrial processes. Fossil fuel emissions increased by 29 percent between 2000 and 2008.

This massive increase is linked to continuing growth in production, and in international trade in goods and services. It also reflects the increased use of coal as a fuel source. The 2009 global economic slowdown resulted in a decrease of CO_2 emissions.

Industrialized nations are beginning to curb their CO_2 emissions by more efficient fuel use, and by using alternative sources. However, the emissions of many newly industrializing countries have increased markedly since 2000, although their emissions per person are still relatively low.

In recent years, coal has overtaken oil as the largest fossil fuel contributor to greenhouse gas emissions. Although petroleum reserves are limited, there are still hundreds of years of coal reserves worldwide, so it could remain a significant source of emissions. China, for example, depends on coal for almost 80 percent of its electricity, and over 90 percent of heat energy. The decisions it takes over future power generation will have a major impact on levels of atmospheric carbon dioxide.

COAL RESERVES

Percentage share of the world's recoverable coal reserves
2008

Total: 861,000 million tonnes

rest of world 25%

USA 28%

India 7%

Russia 18%

Australia 9%

China 13%

FUEL SOURCE

Total annual CO_2 emitted
by type of fuel
2008
million tonnes

- natural gas
- oil
- coal and coal products

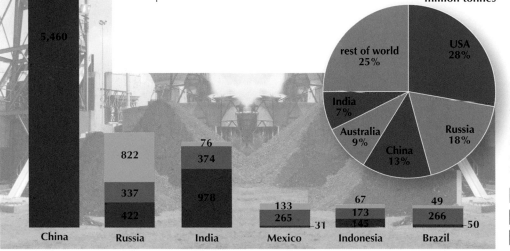

	China	Russia	India	Mexico	Indonesia	Brazil
natural gas	155	822	76	133	67	49
oil	935	337	374	265	173	266
coal and coal products	5,460	422	978	31	145	50

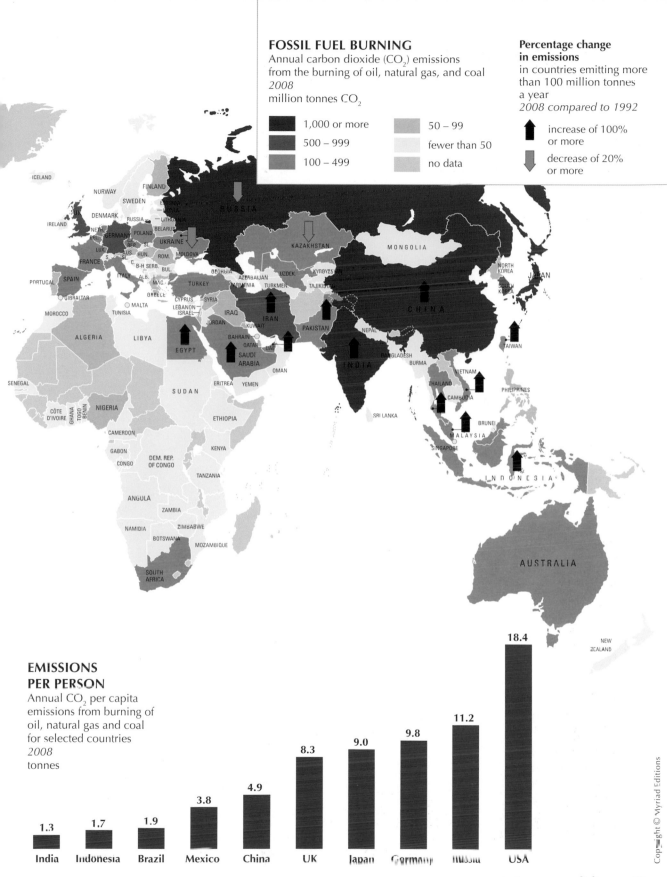

FOSSIL FUEL BURNING

Annual carbon dioxide (CO$_2$) emissions from the burning of oil, natural gas, and coal
2008
million tonnes CO$_2$

- 1,000 or more
- 500 – 999
- 100 – 499
- 50 – 99
- fewer than 50
- no data

Percentage change in emissions

in countries emitting more than 100 million tonnes a year
2008 compared to 1992

- increase of 100% or more
- decrease of 20% or more

EMISSIONS PER PERSON

Annual CO$_2$ per capita emissions from burning of oil, natural gas and coal for selected countries
2008
tonnes

India	1.3
Indonesia	1.7
Brazil	1.9
Mexico	3.8
China	4.9
UK	8.3
Japan	9.0
Germany	9.8
Russia	11.2
USA	18.4

13 METHANE & OTHER GASES

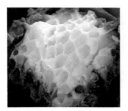

Methane hydrates

An estimated 20 million trillion cubic meters of methane is trapped in permafrost ice and under-sea sediments in a form known as methane hydrates or clathrates. Its release into the atmosphere would be catastrophic, but there is uncertainty about what would trigger a mass release and how much of the gas would be transformed into CO_2 by sea water before it reached the atmosphere. Recent research reports that methane believed to be trapped by under-sea permafrost in part of the Arctic Shelf is being released to the atmosphere.

COMPARATIVE GHG EMISSIONS

Annual emissions including those from land-use change
2005
million tonnes of CO_2e

A range of greenhouse gases contribute to climate change. Methane, nitrous oxide, and other greenhouse gases are more efficient at warming the atmosphere than carbon dioxide, but are present in much smaller quantities, and their overall contribution to global warming is less.

Methane is about 25 times more effective than carbon dioxide at trapping heat in the atmosphere. As its average atmospheric lifetime is 12 years, a reduction in emissions would have a rapid effect. Methane is produced by rice cultivation, coal mining, energy production and the rearing of livestock. In the industrial world, landfill sites are a major contributor. It is also produced in the natural environment, as bacteria break down organic material in anaerobic conditions.

Nitrous oxide is 300 times more effective than carbon dioxide at trapping heat in the atmosphere. The great majority of it is emitted by the agricultural sector – through nitrogen-based fertilizers and livestock manure – with additional releases in waste, industrial processes and energy use.

Manufactured gases such as halocarbons – including chlorofluorocarbons (CFCs), hydrofluorocarbons (HFCs), and perfluorocarbons (PFCs) – and compounds such as sulfur hexafluoride (SF6) have long lifetimes in the atmosphere. Sulfur hexafluoride is used as an insulator for circuit breakers, and to stop oxidation of molten magnesium during processing. HFCs are used in refrigeration units, in place of CFCs.

CANADA

MEXICO

BAHAMAS
CUBA
DOMINICAN REP.
JAMAICA HAITI
ANTIGUA & BARBUDA
BELIZE
ST KITTS & NEVIS
DOMINICA
GUATEMALA HONDURAS
ST VINCENT & GRENAD
ST LUCIA
EL SALVADOR
NICARAGUA
BARBADOS
GRENADA
TRINIDAD & TOBAGO
COSTA RICA
PANAMA
VENEZUELA
GUYANA
COLOMBIA
SURINAME
ECUADOR
PERU
BOLIVIA
PARAGU
CHILE
URUGUAY
ARGENTINA

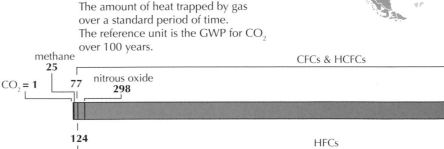

nitrous oxid
3,286

methane
6,408

COMPARATIVE GLOBAL WARMING POTENTIAL

The amount of heat trapped by gas over a standard period of time. The reference unit is the GWP for CO_2 over 100 years.

methane
25

$CO_2 = 1$

77

nitrous oxide
298

CFCs & HCFCs

124

HFCs

 ◀◀ *34–35 The Greenhouse Effect*

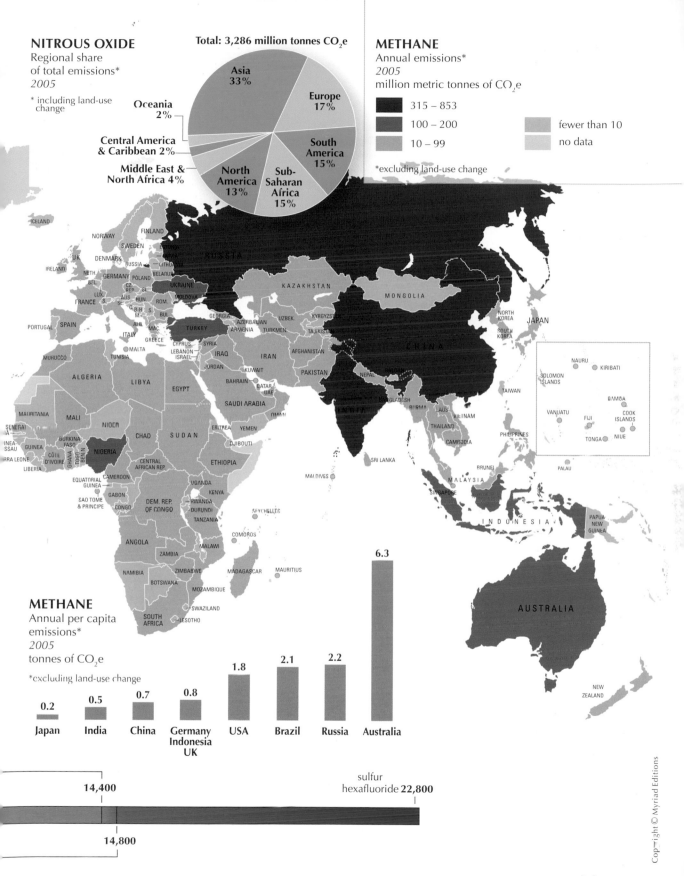

NITROUS OXIDE
Regional share
of total emissions*
2005

* including land-use
change

Total: 3,286 million tonnes CO$_2$e

Asia
33%

Oceania
2%

Central America
& Caribbean 2%

Middle East &
North Africa 4%

North
America
13%

Sub-
Saharan
Africa
15%

South
America
15%

Europe
17%

METHANE
Annual emissions*
2005
million metric tonnes of CO$_2$e

315 – 853	
100 – 200	fewer than 10
10 – 99	no data

*excluding land-use change

METHANE
Annual per capita
emissions*
2005
tonnes of CO$_2$e

*excluding land-use change

0.2	0.5	0.7	0.8	1.8	2.1	2.2	6.3
Japan	India	China	Germany Indonesia UK	USA	Brazil	Russia	Australia

14,400

14,800

sulfur
hexafluoride **22,800**

14 TRANSPORT

Transport accounted for **22%** *of CO_2 emissions in 2008*

International trade and travel, and increasing dependence on motor vehicles make transportation one of the main sources of greenhouse gas emissions.

The growing mobility of both goods and people is the primary force behind a 45 percent increase in emissions from transportation between 1990 and 2007. Passenger and freight travel by road is the largest source of these emissions. Since the 1950s, many nations have become committed to road-based transportation, which is generally less fuel efficient than either rail or shipping. In the USA, 92 percent of households own at least one car, and vehicle miles increased by 1 percent annually between 2000 and 2008. Although there are many options to increase efficiency, heavy investment in road infrastructure makes car dependency difficult to change, and it is an investment decision many rapidly developing countries are currently facing.

Other aspects of lifestyles in industrialized and industrializing nations – the desire to travel widely, and the expectation of a global supply of goods and food – are also contributing to increased transport emissions. Air travel, once accessible only to the wealthy, has become more affordable to many. In the absence of new policies, the International Energy Agency projects a nearly four-fold increase in passenger air travel and air freight between 2005 and 2050. Emissions from international shipping doubled between 1987 and 2007, reflecting increased trade in manufactured goods, food, and raw materials.

Emissions from international shipping and from aircraft are not attributed to individual nations under the Kyoto Protocol, and there is no global target for their reduction.

42% of road emissions are produced in the USA

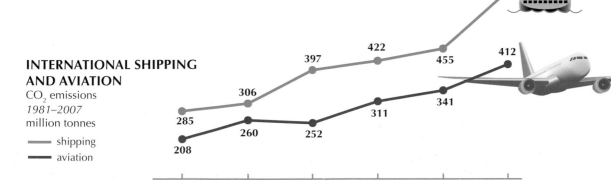

INTERNATIONAL SHIPPING AND AVIATION
CO_2 emissions
1981–2007
million tonnes
— shipping
— aviation

shipping: 285, 306, 397, 422, 455, 610
aviation: 208, 260, 252, 311, 341, 412

1981 1987 1992 1997 2002 2007

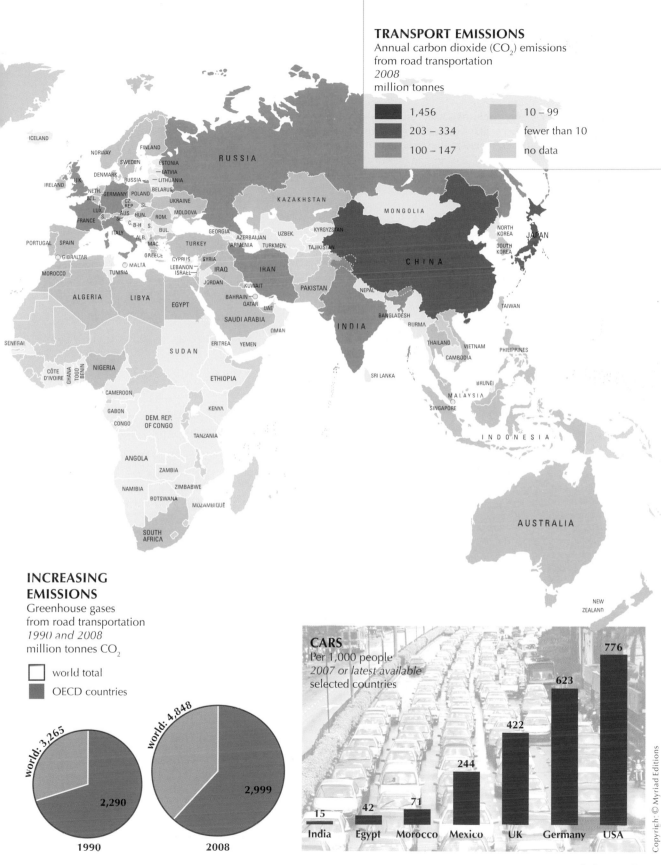

TRANSPORT EMISSIONS

Annual carbon dioxide (CO$_2$) emissions
from road transportation
2008
million tonnes

- 1,456
- 203 – 334
- 100 – 147
- 10 – 99
- fewer than 10
- no data

INCREASING EMISSIONS

Greenhouse gases
from road transportation
1990 and 2008
million tonnes CO$_2$

- ☐ world total
- ▨ OECD countries

world: 3,265

2,290

1990

world: 4,848

2,999

2008

CARS

Per 1,000 people
2007 or latest available
selected countries

India	Egypt	Morocco	Mexico	UK	Germany	USA
15	42	71	244	422	623	776

15 AGRICULTURE

The luxury of choice
Wealthy consumers, in industrialized countries, and increasingly in developing countries too, expect an ever wider choice of food products. Agriculture worldwide is responding to consumer demands for a year-round choice of fresh fruit and vegetables. The greenhouse gases released in bringing food products to our tables, by air, sea, and road, are increasingly being considered in calculations of the carbon footprint of people in industrialized economies.

Agriculture is a major source of greenhouse gases, and vital for economic welfare in many countries. It accounts for about one-third of global emissions of carbon dioxide, methane, and nitrous oxide.

In wealthier countries, agriculture is primarily a commercial activity, accounting for a relatively small proportion of the economy. In many developing countries, however, it is the main economic activity of the rural population – representing between 25 percent and 50 percent of GDP. It is essential to meet basic needs: food, employment, and income. Changes in consumer demand and trade, and fluctuations in prices for energy and fertilizers, can be significant. In countries that depend on food imports – such as North Africa and the Middle East – rising global food prices have significant impacts on large sections of the population.

Plans for reducing emissions in agriculture must consider the consequences for less advantaged populations. The growing of rice in flooded fields releases methane from waterlogged soils, but rice feeds a third of the world's population and is the staple diet of many poor people in Asia. Livestock are also a source of methane, but those on the small holdings of poor farmers and pastoralists produce much less gas than the well-fed cattle in large commercial enterprises. Fertilizers produce nitrous oxide, but also provide a much-needed boost to food production in some areas.

COMPARATIVE PROFILES

Distribution of agricultural emissions throughout agri-food chain
2005

- production
- processing
- distribution and retail
- consumption

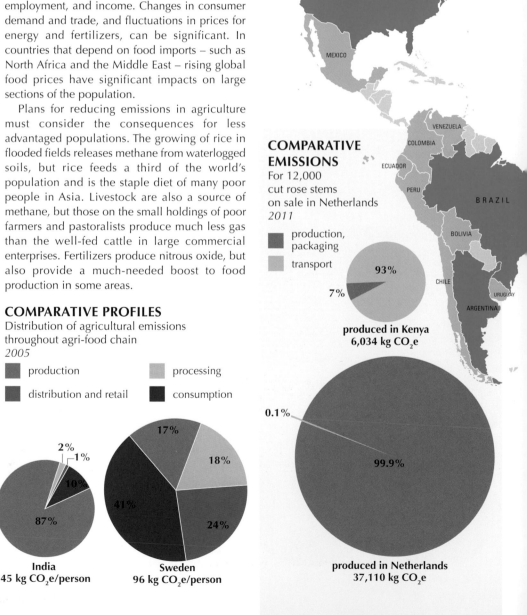

India
45 kg CO_2e/person

2%
1%
10%
87%

Sweden
96 kg CO_2e/person

17%
18%
41%
24%

COMPARATIVE EMISSIONS

For 12,000 cut rose stems on sale in Netherlands *2011*

- production, packaging
- transport

produced in Kenya
6,034 kg CO_2e

93%
7%

produced in Netherlands
37,110 kg CO_2e

99.9%
0.1%

CANADA
USA
MEXICO
VENEZUELA
COLOMBIA
ECUADOR
PERU
BRAZIL
BOLIVIA
CHILE
URUGUAY
ARGENTINA

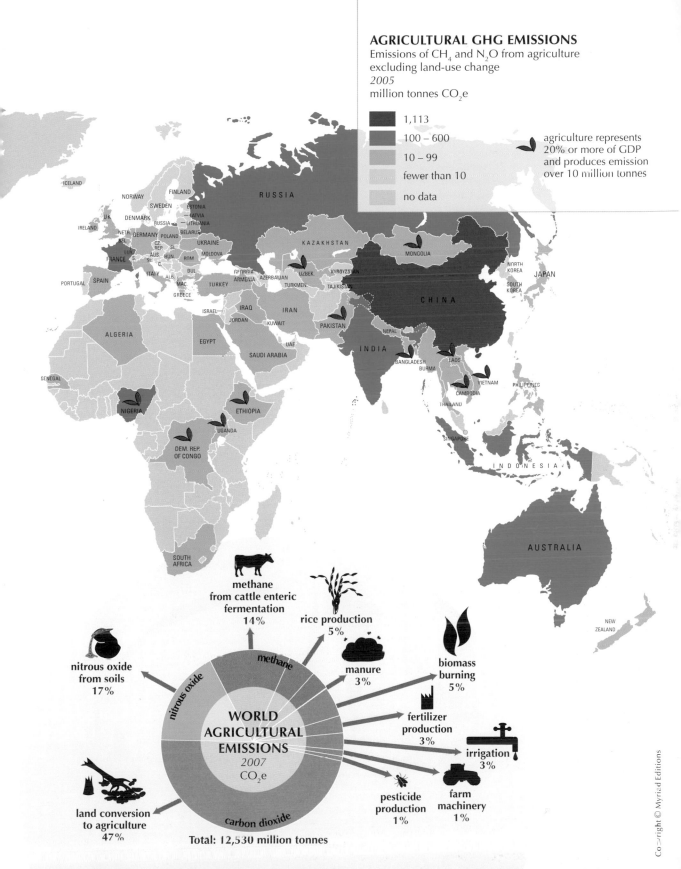

AGRICULTURAL GHG EMISSIONS

Emissions of CH_4 and N_2O from agriculture excluding land-use change
2005
million tonnes CO_2e

- 1,113
- 100 – 600
- 10 – 99
- fewer than 10
- no data

agriculture represents 20% or more of GDP and produces emission over 10 million tonnes

WORLD AGRICULTURAL EMISSIONS
2007
CO_2e

nitrous oxide from soils
17%

land conversion to agriculture
47%

methane from cattle enteric fermentation
14%

rice production
5%

manure
3%

biomass burning
5%

fertilizer production
3%

irrigation
3%

pesticide production
1%

farm machinery
1%

Total: 12,530 million tonnes

16 THE CARBON BALANCE

By 2030 we will need the capacity of **2 Earths** *to absorb CO$_2$ and provide natural resources*

Carbon is plentiful in the natural environment, stored in plants and soils, and dissolved in the oceans. But changes in stored carbon are contributing to climate change, which may, in turn, be accelerating the release of carbon.

The absorption of carbon from the atmosphere by plants and soil organisms, and its release through waste and decay, is part of the natural carbon cycle. When people cut down forests to make way for intensive cultivation, or for buildings, more carbon is released than is absorbed. The natural carbon balance is disrupted.

About a quarter of the carbon released into the atmosphere over the last 150 years has arisen from a change in the way land is used. During the 20th century and, in particular, since the 1950s, deforestation in South America, Africa, and parts of Asia has released large amounts of carbon, as have farming practices in dry regions, such as in west and east Africa and in India. Over the same period, however, there was much less change in land use in North America and Europe, where forests and tree plantations serve as carbon sinks on a modest scale.

The contribution of land-use change to climate change has not altered much in the past decade, and further action on reducing carbon emissions from terrestrial land-use change is expected through the REDD+ strategy (reducing emissions from deforestation and land degradation).

Oceans also exchange carbon dioxide with the atmosphere, and store a vast amount of carbon in their deeper layers. Carbon embodied in dead plants and animals sinks to the ocean floor, and is thus removed from the atmosphere. If oceans become significantly warmer, they may start to release more carbon than they absorb.

The combination of climate change and land-use change can destabilize large carbon reservoirs and reduce the efficiency of natural CO$_2$ sinks. Vulnerable reservoirs include oceans, carbon in frozen soils and northern peat, tropical peat, forests vulnerable to deforestation, drought and wildfires, and methane hydrates in permafrost and continental shelves. The efficiency of CO$_2$ uptake by natural sinks may have already declined over the last 50 years.

CARBON CYCLE
Carbon stores and annual fluxes in million tonnes *2000–09*

stores

flows

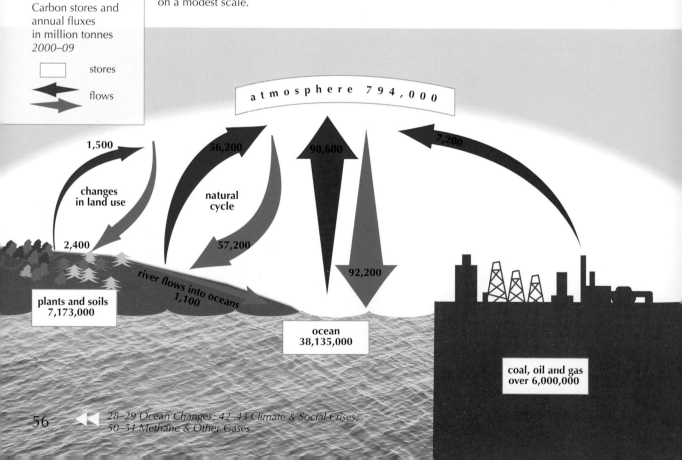

atmosphere 794,000

1,500

changes in land use

2,400

56,200

natural cycle

57,200

90,600

92,200

7,200

river flows into oceans 1,100

plants and soils 7,173,000

ocean 38,135,000

coal, oil and gas over 6,000,000

LAND-USE CHANGE

Net carbon emitted or absorbed into the atmosphere
from deforestation, shifting cultivation, and vegetation
re-growth on abandoned croplands and pastures
1950–2000
million tonnes carbon

carbon emitted

- more than 5,000
- 1,500 – 5,000
- up to 1,500
- no data

carbon absorbed

- up to 100
- 100 – 500
- more than 500

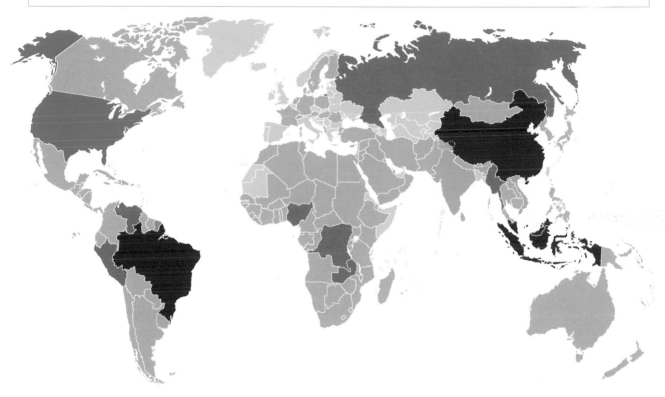

TRENDS IN CARBON SOURCES AND SINKS
Annual emission and absorption of carbon
1850–1990

- emissions from fossil fuels
- net release from land-use change
- remaining terrestrial sink
- oceanic uptake
- atmospheric increase

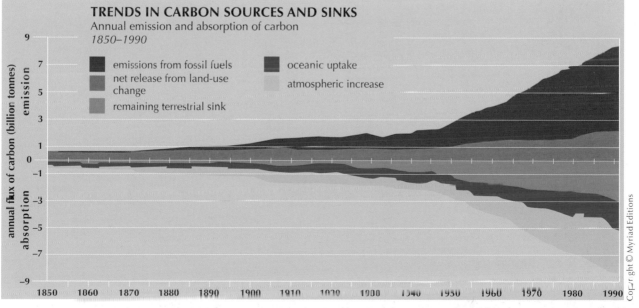

PART 4 Expected Consequences

Our current vulnerabilities to climate, and the way those stresses are compounded or ameliorated by other life circumstances, is the basis for understanding the potential consequences of projected climate changes. Warming trends are affecting ecosystems and resources in unprecedented ways. Approximately 1.4 billion people live in poverty, some of them dependent on agriculture, and most lacking access to clean water or basic health care. Many live in regions where water and food security are expected to be further stressed by climate changes. The most vulnerable people are likely to be hit earlier and harder than others.

Wealth, however, will not provide shelter from all impacts. In recent years, residents of rich and poor countries alike have been displaced by catastrophic flooding. Wildfires have taken homes in wealthy as well as poorer neighborhoods. Even those spared from disaster are likely to feel the pressure through the global connectedness of markets.

The confidence of the international scientific community that what we are seeing is the impact of climate change is continuing to grow; the increased frequency of extreme weather events has been attributed to the increased concentrations of greenhouse gases in the atmosphere. The consequences of climate change will extend everywhere through both direct and indirect paths. Global trade and information flows connect localities around the world. The impacts on agriculture may begin with changes in crop yields in a field, but may ultimately be felt in higher prices in cities or countries on the other side of the world.

Sea-level rise and reduced sea-ice coverage in the Arctic offer another example of interconnectedness through both physical and social systems. Ports represent hundreds of millions of dollars in infrastructure and draw in a variety of industries and employment to a region. Currently they are in competition for the associated economic development opportunities. If warming in the Arctic were to open a new, year-round shipping route, competitive advantages among ports would change. In the meantime, all port facilities will have to adjust to rising sea levels. In some developing countries, where ports generate the foreign currency needed to pay national debts and to foster further economic development, making those improvements will be crucial and difficult to afford.

The diversion of national funds to respond to disasters and to maintain key economic resources competes with urgent demands for health care, education, and other basic needs. Climate change is not solely an environmental issue; it has significant implications for our ability to achieve broader societal goals.

> The security of people and nations rests on four pillars – food, energy, water and climate. They are all closely related, and all under increasing stress.
>
> **Tom Burke**
> former CEO of Friends of the Earth, founder of E3G, and environmental policy advisor to Rio Tinto, 2008

17 Disrupted Ecosystems

1.5°C to 2.5°C warming will put up to

30%

of species studied at high risk of extinction

Many species and ecosystems, already under stress from human development, may not be able to adapt to new climatic conditions and pressures.

Some of the regions richest in biodiversity are already being affected, while also coping with the pressures of other human uses, such as forestry and agriculture. Catastrophic events, such as droughts, changing frequency of fire or insect outbreaks, or even small changes in average temperature, can disrupt ecosystems built on the inter-dependency of thousands of species. Sea-level rise is destroying low-lying habitats which play important roles in ocean, estuarine, and terrestrial ecosystems.

Unprecedented rates of migration by both plants and animals will be needed if species are to keep up with climate change. While grasslands and desert can spread fairly quickly into new areas, slower-growing forests may find themselves outpaced in the race against time. Interactions among environmental changes may result in abrupt transitions that make it more difficult to respond. Some impacts will be immediately apparent, while slow processes, such as soil formation, mean that some ecosystem adaptations will continue over centuries.

Not all systems are equally well positioned to allow changes in range. In mountainous regions, some species may only have to move a few hundred meters up hill to find new terrain, but limits to ecosystem-based adaptation arise when those already at the top of the mountain or on the most northerly landmasses in the Arctic need to move to cooler climes. Some species have their escape routes blocked off, surrounded as they are by agricultural and urban development. The creation of migration corridors is a focus of much conservation work.

The annual destruction of 5 million hectares of humid tropical forest is of particular concern. The Earth's forests play a role in the natural process of storing carbon, while also providing habitat protection. Forest destruction accelerates both the impacts of climate change and the loss of biodiversity.

> Clearly, we are endangering all species on Earth. We are endangering the future of the human race.

Rajendra Pachauri
IPCC Chair, 2007

Non-native species

Climate change is expected to make ecosystems susceptible to invasion by non-native species in circumstances when new species are favored by changing habitat conditions.

Prairie wetlands

Projected higher temperatures threaten prairie pothole wetlands, a key habitat in which waterfowl such as mallard and canvasback (pochard) duck raise their young.

Brazilian Cerrado

This savanna (grassland) originally covered more than 20% of Brazil. Now it is so fragmented that it is unlikely that its 10,000 plant species will all be able to disperse to climatically suitable areas.

Adélie penguins

Changing conditions on the Antarctic Peninsula are contributing to the decline of **Adélie penguin** populations.

Changes consistent with climate change

90% of approximately 28,800 documented changes in biological systems between 1970 and 2004 are consistent with scientists' expectations of the effects of climate change. These include shifts in spring events, such as blooming date, and time of reproduction, as well as other impacts including changes in species distributions.

Europe

The ability of European birds and plants to migrate in the face of warmer temperatures is severely limited by the density of the human population, and by intensive cultivation. The Spanish Imperial Eagle, for example, currently largely restricted to nature parks and reserves, may find that unspoilt areas with suitable temperatures are not available. The Scottish Crossbill, endemic to Scotland, UK, might be faced with a move to Iceland – a journey it would be unlikely to undertake successfully.

BIODIVERSITY AT RISK

Ecosystems and species threatened by climate change
2010

Sundarbans Delta

This huge area of mangroves in the Bay of Bengal, home to the **Bengal tiger** and other important species, is threatened by rising sea levels.

At least 40% of the world's economy and 80% of the needs of the poor are derived from biological resources.

Secretariat of the Convention on Biological Diversity

Australia

Climate change, coupled with the stresses of land-use changes, is likely to result in substantial declines in habitat suitable for **banksia**, a genus of approximately 80 species of showy, nectar-providing plants almost entirely endemic to Australia.

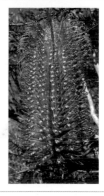

Mount Kilimanjaro

On Mt. Kilimanjaro, glacial retreat resulting in drier conditions and increased fire frequency has contributed to a reduction of forest cover and an increase in shrub land.

Ocean warming

By allowing the spread of disease agents and parasites, ocean warming may contribute to the outbreak of species-based epidemics, such as the one that killed thousands of **striped dolphins** in the Mediterranean in the early 1990s.

GLOBAL WARMING

Areas where ecosystems will change, for a scenario of a global mean temperature increase of 3°C.

The effect of warmer temperatures will lead to a general shift of ecosystems towards the poles. In some areas there is no available land at higher latitude, which will probably lead to the disappearance of an ecosystem.

- ice and tundra
- woodland and forest
- grassland and scrub
- desert

Map courtesy of Rik Leemans, Wageningen University, Netherlands.

Copyright © Myriad Editions

18 WATER SECURITY

Nearly
3 billion
*people live
in areas
where water
demand
outstrips
supply*

Up to
5 billion
*people may
live in
water-stressed
areas
by 2050*

Water scarcity has already become a major stress in some regions. Climate change may raise the stress level in some places, which increases the urgent need for global action.

Water is a vital resource, often taken for granted, but rising demand from expanding populations, and an increasing risk of drought are causing concern in many countries. It is now apparent that climate change is making the situation worse.

Higher temperatures result in more evaporation from surface water, and more evapotranspiration from plants, reducing supply and increasing demand, especially for irrigation water. Warmer and longer summers cause snow packs and glaciers to melt more quickly, which increases river flows in the spring, but may reduce summer flows. Over the long term, a reduction in snow and ice may seriously threaten many river basins. For example, in northern India 500 million people rely on the Indus and Ganges, which are fed largely by glacial melt waters.

Some areas will experience less annual rainfall; in others it will be less predictable, with seasonal rains failing to materialize, or arriving with such ferocity that they create dangerous floods. Other threats to freshwater supplies include sea-level rise, which leads to saline intrusion in coastal aquifers, and damage to water infrastructure from coast storms.

Rapid changes in weather patterns – from seasonal crises to a decade of low flows – leave people little time to adapt. Regions reliant on heavily irrigated agriculture may be forced to import more food, or even to import water itself. International water imports have been suggested as urgent options in the Mediterranean and southeast England. Water-intensive industries, such as paper and electronics manufacturing, will be unable to function, and economies will suffer as a consequence. If water supplies fail completely, contaminated water, lack of hygiene and thirst will take their toll.

Many of the effects of climate change can be countered by prioritizing the most urgent uses, adopting water-saving technology and more efficient irrigation methods. The cost of adaptation in the water sector in developing countries has been estimated as up to $20 billion a year. However, the finance, technology, and infrastructure necessary to effectively manage current water resources are significant barriers.

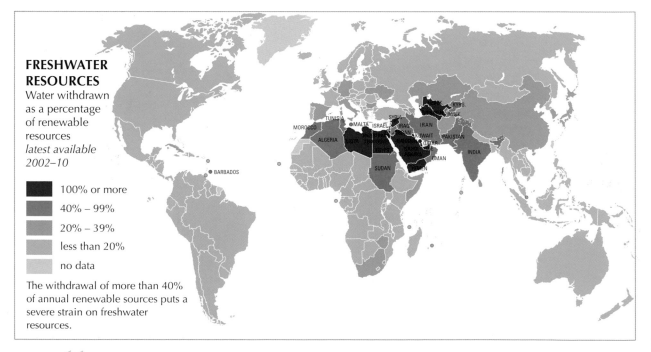

FRESHWATER RESOURCES
Water withdrawn as a percentage of renewable resources
latest available 2002–10

- 100% or more
- 40% – 99%
- 20% – 39%
- less than 20%
- no data

The withdrawal of more than 40% of annual renewable sources puts a severe strain on freshwater resources.

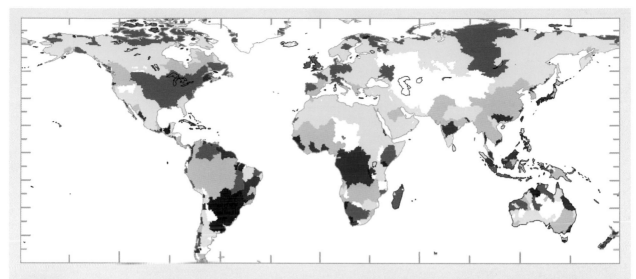

TREND IN RIVER RUN-OFF
Millimeters
1948–2004

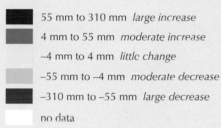

- 55 mm to 310 mm *large increase*
- 4 mm to 55 mm *moderate increase*
- –4 mm to 4 mm *little change*
- –55 mm to –4 mm *moderate decrease*
- –310 mm to –55 mm *large decrease*
- no data

Some impacts of climate change have already been observed. The annual trend in run-off (the balance of rainfall and evapotranspiration) in major river basins over the past 50 years may be an early indicator of future climate impacts. The pattern is mixed across the world, although few areas show large increases.

Run-off decreases are strongly related to warmer temperatures and decreases in precipitation; both are linked to episodes of the El Niño/Southern Oscillation. However, it is too early to fully attribute such trends to climate change.

Water planning often looks forward beyond 30 years to evaluate a range of potential futures. Increasing aridity in the Mediterranean is likely, in contrast to increased run-off in northern temperate regions. However, in large parts of the world, the balance of change is unclear.

POTENTIAL CHANGE IN RUN-OFF
Possible large-scale changes in annual water run-off
2090–99 relative to 1980–99

Decrease
- 40% or more
- 20% – 39%
- 10% – 19%
- 5% – 9%
- 2% – 4%
- little change

Increase
- 2% – 4%
- 5% – 9%
- 10% – 19%
- 20% – 39%
- 40% or more
- fewer than 66% of models agree
- more than 90% of models agree

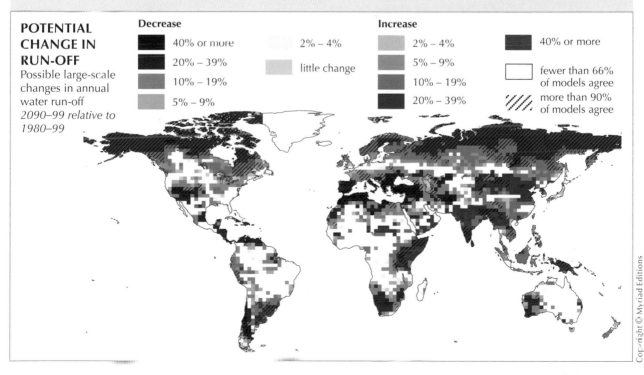

19 FOOD SECURITY

Investment of

$7 billion

a year is required to reduce climate impacts on agriculture in developing countries

Food security continues to plague vulnerable populations. Climate change threatens future agricultural development in many regions, although it is benefiting crop production in some temperate regions.

The effect of climate change on agriculture depends on a combination of factors. Higher temperatures can stress plants, but also prolong growing seasons and allow a greater choice of crops to be grown. Higher concentrations of carbon dioxide speed growth and increase resilience to water stress. Much depends on rainfall within the growing season, and on irrigation. Pests and diseases may also increase in response to more benign climatic conditions.

Globally, climate change already appears to have affected crop production, although the overall impact has been more significant for some crops than for others. Regional projections to 2030, based on climate models, reveal a complex picture.

Agriculture in temperate climates may actually benefit from longer growing seasons and warmer temperatures. Elsewhere, farmers will adapt, altering their crop calendars to avoid extreme hot periods, introducing plant varieties that can tolerate a range of conditions, and using good soil management to overcome water stress. World food production should be able to adjust to climate change through 2050 or so, provided there is significant investment in agricultural and economic development.

In parts of the tropical and sub-tropical regions, however, reductions in rainfall, and increasing risk of drought, or of more intense rainfall and soil erosion, will severely affect agriculture. The capacity of developing countries to sustain agricultural production and food security is already challenged. The poor, those most likely to experience malnutrition, are likely to suffer further.

MALNOURISHED CHILDREN

Estimated numbers under varying circumstances in 2050

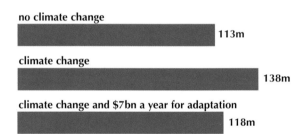

no climate change
113m

climate change
138m

climate change and $7bn a year for adaptation
118m

ESCALATING FOOD PRICES

Global cereals price index (2002–04 = 100)
2000–11

The global food crisis of 2007–08 was followed by even higher prices in late 2010 and early 2011. The rising prices were attributed to conversion of cereals to biofuels, rising energy prices, reduced harvests related to droughts, floods and fires, and speculation. High food prices were one factor in the riots and revolutions in North Africa and the Middle East in 2011: the so-called Arab Spring.

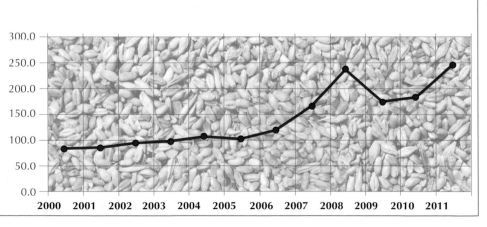

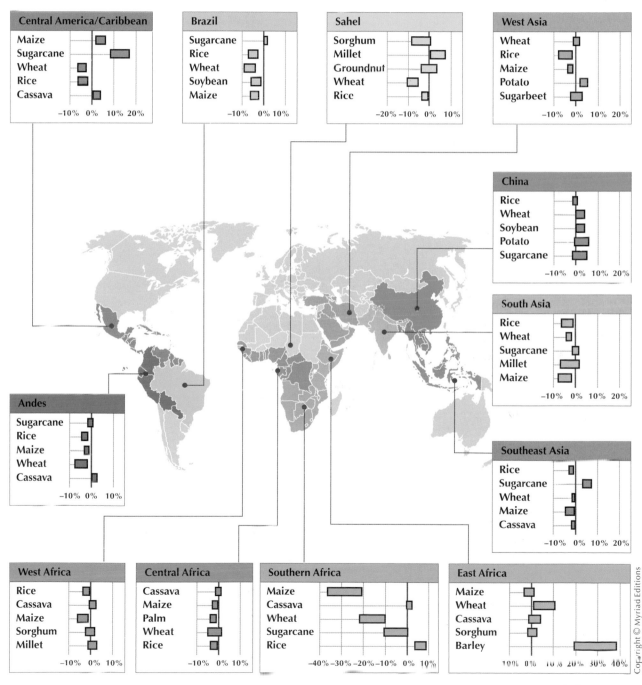

IMPACT OF CLIMATE CHANGE ON CROP PRODUCTION

Projected yield for 2030
as percentage of average yield
for 1998–2002

 mid-range projections for
most nutritionally significant
crops in each region

Central America/Caribbean
Maize
Sugarcane
Wheat
Rice
Cassava

–10% 0% 10% 20%

Brazil
Sugarcane
Rice
Wheat
Soybean
Maize

–10% 0% 10%

Sahel
Sorghum
Millet
Groundnut
Wheat
Rice

–20% –10% 0% 10%

West Asia
Wheat
Rice
Maize
Potato
Sugarbeet

–10% 0% 10% 20%

China
Rice
Wheat
Soybean
Potato
Sugarcane

–10% 0% 10% 20%

South Asia
Rice
Wheat
Sugarcane
Millet
Maize

–10% 0% 10% 20%

Andes
Sugarcane
Rice
Maize
Wheat
Cassava

–10% 0% 10%

Southeast Asia
Rice
Sugarcane
Wheat
Maize
Cassava

–10% 0% 10% 20%

West Africa
Rice
Cassava
Maize
Sorghum
Millet

–10% 0% 10%

Central Africa
Cassava
Maize
Palm
Wheat
Rice

–10% 0% 10%

Southern Africa
Maize
Cassava
Wheat
Sugarcane
Rice

–40% –30% –20% –10% 0% 10%

East Africa
Maize
Wheat
Cassava
Sorghum
Barley

–10% 0% 10% 20% 30% 40%

20 THREATS TO HEALTH

Over **70,000** *people died in weather-related disasters in 2010*

Under-nutrition is the underlying cause of death for at least **30%** *of all under-fives*

People already suffer from a variety of weather-related health effects. The number of victims and the economic costs have increased in the past decade, possibly a sign of future threats to health.

Exposure to weather-related disasters, such as heat waves and cyclones, and changes in resources, such as sea-level rise, ocean acidification, and altered growing seasons, means everyone is potentially at risk of one health consequence or another as a result of climate change.

Diseases transmitted by vectors, such as mosquitoes and ticks, and pests that destroy crops, are affected by rising temperatures and humidity, and by altered rainfall patterns. The areas affected by diseases may expand in some regions, contract in others, or the disease may become more common throughout the year. The absence of pest-killing sub-zero temperatures increases year-round populations of pests.

While fewer people may die from cold, warmer weather leads to increased heat stress. It may also lead to higher levels of air pollutants from forest fires in rural areas, and from the formation of ozone and volatile organic compounds in urban areas. The number of deaths related to respiratory conditions will rise.

Intense rainfall and flooding increases the risk of waterborne diseases such as cholera, typhoid, and dysentery, and of mosquito-borne diseases, including malaria and yellow fever. Heavy rainfall often leads to contamination of the environment and water supplies: poor sanitation is a major health risk in crowded urban areas.

Drought and disasters reduce food supplies, with direct effects on malnutrition in vulnerable people. Poor nutrition is a significant factor in the ability to fight off infections and reduces the effectiveness of many medical treatments (including for HIV/AIDS).

The effects of disasters and changing environments on mental health are a growing concern. Increased suicide rates in drought-affected areas have been noted. Anxiety has been caused by new risks such as coastal erosion and the need to relocate communities.

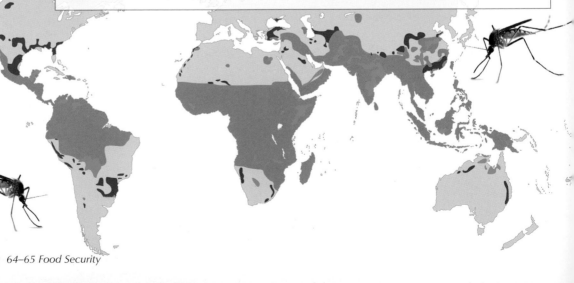

POTENTIAL FUTURE MALARIA RISK
Scenario of malaria distribution
projected 2050

- current distribution of falciparum malaria
- 2050 climate suitable for vector, so malaria may appear
- 2050 climate unsuitable for vector, so malaria may disappear

The map shows the distribution of the parasite that causes malaria in humans: *Plasmodium falciparum*, carried by the *Anopheles* mosquito. It is currently distributed wherever conditions are favorable. Areas of expansion and contraction are based on one climate scenario, as an indication of the potential change in risk. The areas are roughly equal in extent, and include about the same number of people: 400 million.

INCREASED DISEASE BURDEN
Number of extra cases of diarrhea, malaria, and malnutrition,
assuming a business-as-usual scenario of climate change
projected for 2030

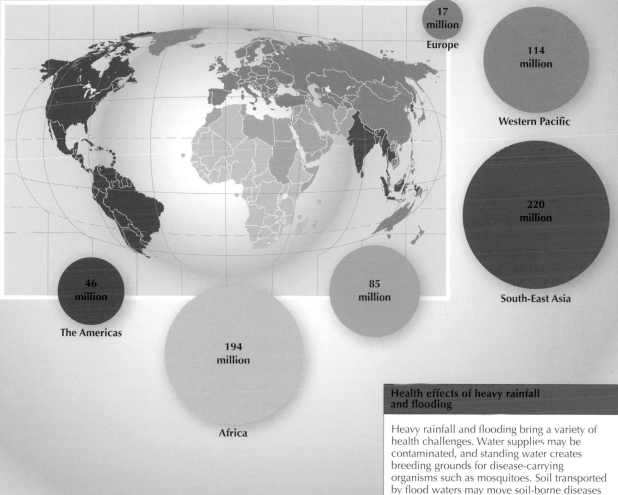

17 million
Europe

114 million
Western Pacific

220 million
South-East Asia

46 million
The Americas

85 million

194 million
Africa

Tick-borne diseases

Lyme disease, which can cause long-term disability, is spreading in the USA and Europe as winter temperatures warm and daily temperatures rise. The habitat favorable to the **deer ticks** that carry the disease is likely to spread. Rocky Mountain spotted fever, Q fever, and, especially in Europe, tick-borne encephalitis are among other diseases that may also spread as the climate changes.

Health effects of heavy rainfall and flooding

Heavy rainfall and flooding bring a variety of health challenges. Water supplies may be contaminated, and standing water creates breeding grounds for disease-carrying organisms such as mosquitoes. Soil transported by flood waters may move soil-borne diseases such as anthrax, and toxic contaminants such as heavy metals and organic chemicals, to previously unexposed areas. Mold developing on flooded property may contribute to respiratory problems. Cholera is related to wetter seasons.

21 RISING SEA LEVELS

WETLANDS LOSS

From sea-level rise in worst-affected countries
by 2100

USA 72,380 km²

Australia 29,830 km²

Indonesia 20,620 km²

Brazil 17,800 km²

Canada 12,490 km²

Mexico 11,590 km²

Chile 11,400 km²

Cuba 11,000 km²

Thermal expansion of oceans and melting ice leads to rising sea levels, threatening many coastal communities.

Mean sea level rose by around 15 centimeters during the 20th century, and projections indicate a further rise of at least 18 centimeters between 1990 and 2100. Even if greenhouse gas emissions are radically reduced over the next 20 years, because of the huge thermal mass of the oceans, sea levels will continue to rise for centuries – a long-term consequence of emissions already released.

The upper range of potential sea-level rise is quite uncertain. Conventional projections show an upper level of just over 0.5 meters by 2100. However, sea levels are rising faster than some of these scenarios and with potential large scale releases from Antarctica and Greenland, by 2100 sea levels might be 2 meters higher than in 1990. Coastal movement – sinking or rising – also affects the height of the sea relative to the land. And storm surges can add 5 meters or more to the local mean sea level.

A rise in global mean sea level of 1 meter will have drastic consequences for many coastal communities. The Maldives islands in the Indian Ocean will be almost completely inundated, as will large parts of island groups in the Caribbean and Pacific. Around the world, valuable agricultural land will be lost, and cities will be threatened. With stronger windstorms possible, many low-lying communities will be at risk from storm surges. The movement of seawater higher up rivers, and into freshwater aquifers, will affect drinking water supplies across the world, threatening the viability of many communities.

Far more serious sea-level rise is possible. Were the Greenland ice cap to melt, it would add an estimated 7 meters to the global sea level. The ice sheet covering West Antarctica rests on rock that is below sea level. Were it to collapse, the sea level might rise by a further 5 meters. At present, there is only a low probability that these ice sheets will collapse in the next few centuries. However, if global warming exceeds 2°C or so, collapse of major ice sheets is more likely, with extreme consequences in all coastal areas.

Under a 1-meter rise in sea level Egypt's GDP could decrease by **6%**

NILE DELTA

▉ area inundated by a 1-meter sea-level rise

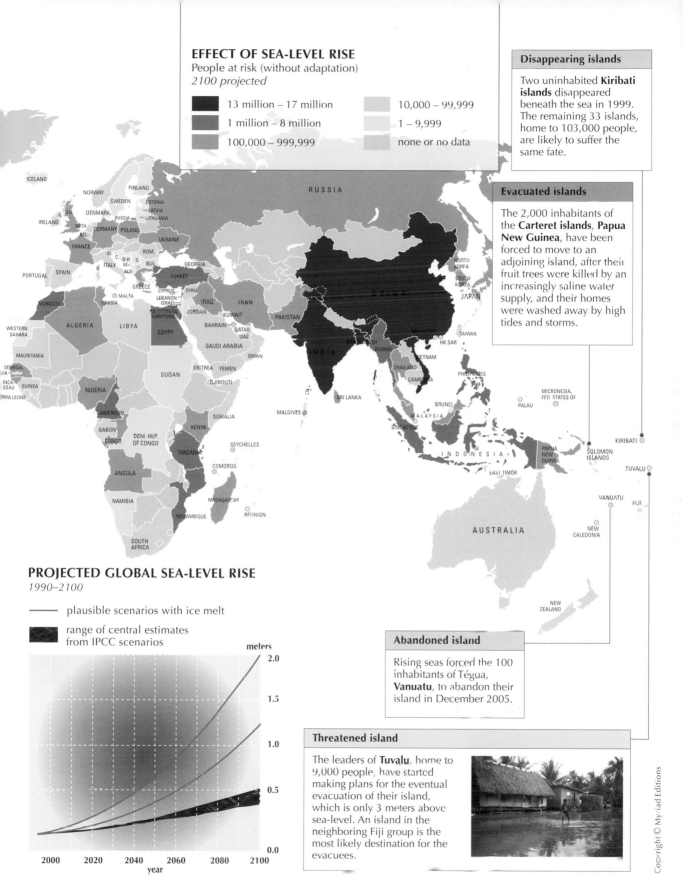

EFFECT OF SEA-LEVEL RISE

People at risk (without adaptation)
2100 projected

- 13 million – 17 million
- 1 million – 8 million
- 100,000 – 999,999
- 10,000 – 99,999
- 1 – 9,999
- none or no data

Disappearing islands

Two uninhabited **Kiribati islands** disappeared beneath the sea in 1999. The remaining 33 islands, home to 103,000 people, are likely to suffer the same fate.

Evacuated islands

The 2,000 inhabitants of the **Carteret islands**, **Papua New Guinea**, have been forced to move to an adjoining island, after their fruit trees were killed by an increasingly saline water supply, and their homes were washed away by high tides and storms.

PROJECTED GLOBAL SEA-LEVEL RISE

1990–2100

— plausible scenarios with ice melt

■ range of central estimates from IPCC scenarios

meters

2.0
1.5
1.0
0.5
0.0

2000 2020 2040 2060 2080 2100
year

Abandoned island

Rising seas forced the 100 inhabitants of Tégua, **Vanuatu**, to abandon their island in December 2005.

Threatened island

The leaders of **Tuvalu**, home to 9,000 people, have started making plans for the eventual evacuation of their island, which is only 3 meters above sea-level. An island in the neighboring Fiji group is the most likely destination for the evacuees.

22 CITIES AT RISK

Almost **22,000** *cities of over 100,000 people are in coastal zones*

Coastal storms, flooding, inundation, erosion, and saltwater intrusion into freshwater supplies present a combined threat to coastal areas.

Around 40 percent of people live less than 60 miles from the coast – within reach of severe coastal storms. About 145 million people live less than one meter above mean sea level. Economic activities concentrated in larger coastal cities continue to draw in people seeking jobs and opportunities. Although urban areas have more resources than rural areas, they face the threat of huge losses.

Coastal erosion, rising sea levels, saltwater contamination, and potentially more powerful storms, are expected to put these low-lying urban environments under increasing stress. Aging infrastructure may be inadequate under current circumstances or become (more) inadequate under future climate conditions. Communication, trade, healthcare, and transportation services, also necessary to the well-being of a city, are at risk.

Some of the consequences of climate change, such as the inundation of large delta areas, are potentially catastrophic. Others, such as the movement of saltwater upstream into freshwater rivers, flooding at high tide, or unusually heavy rainstorms, are expected to take their toll through increased frequency or gradual increases over time. Saltwater intrusion can cause drinking and irrigation water to become saline, river water to becomes too corrosive to use for cooling in industrial processes and power plants, and to change coastal ecosystems.

While all coastal cities face such threats, those with over 10 million inhabitants will face particularly steep challenges due to their greater scale and complexity, and the enormous and diverse populations they serve. Water and sanitation systems may be placed under unbearable strain, and millions of poor people in shanty towns on the fringes of the cities may be at even greater risk from disease. Port facilities may no longer be viable, and government and financial services may be severely damaged, affecting the administration and economy of the entire country.

Port cities
There are 136 port cities with populations over 1 million. By 2070, the total value of the assets threatened by a climate change scenario of 0.5 meters of sea-level rise, increased storminess, land subsidence, population growth, urbanization, and economic growth could reach $35 trillion – ten times current assets, and roughly 9% of projected GDP.

New Orleans

The effect of a powerful hurricane on a low-lying city was demonstrated by Hurricane Katrina in August 2005. The storm surge it created flooded large areas of New Orleans, and caused damage valued at over $50 billion.

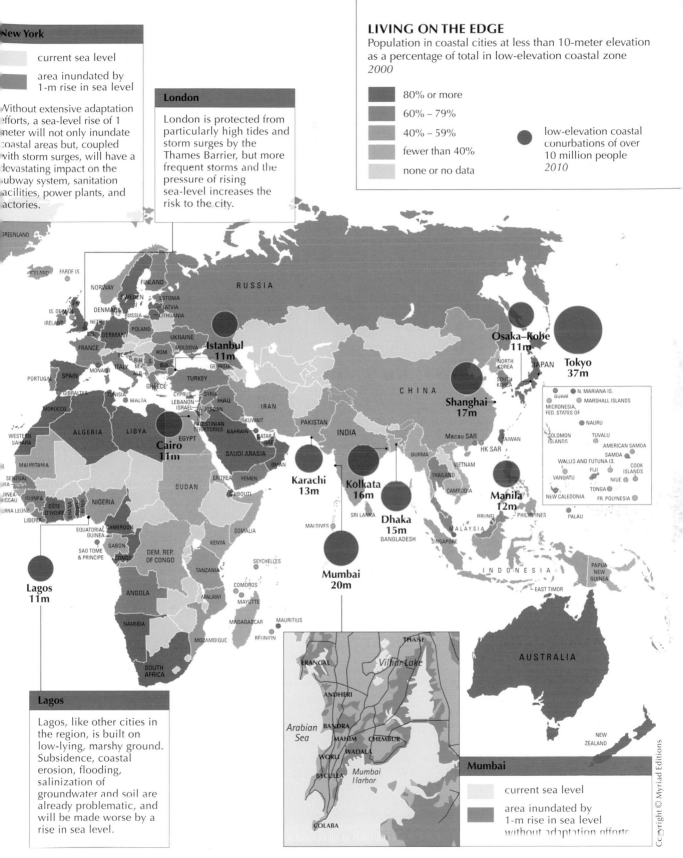

LIVING ON THE EDGE

Population in coastal cities at less than 10-meter elevation as a percentage of total in low-elevation coastal zone
2000

- 80% or more
- 60% – 79%
- 40% – 59%
- fewer than 40%
- none or no data

low-elevation coastal conurbations of over 10 million people
2010

New York

current sea level

area inundated by 1-m rise in sea level

Without extensive adaptation efforts, a sea-level rise of 1 meter will not only inundate coastal areas but, coupled with storm surges, will have a devastating impact on the subway system, sanitation facilities, power plants, and factories.

London

London is protected from particularly high tides and storm surges by the Thames Barrier, but more frequent storms and the pressure of rising sea-level increases the risk to the city.

Lagos

Lagos, like other cities in the region, is built on low-lying, marshy ground. Subsidence, coastal erosion, flooding, salinization of groundwater and soil are already problematic, and will be made worse by a rise in sea level.

Mumbai

current sea level

area inundated by 1-m rise in sea level without adaptation efforts

Istanbul 11m
Osaka–Kobe 11m
Tokyo 37m
Shanghai 17m
Cairo 11m
Karachi 13m
Kolkata 16m
Manila 12m
Dhaka 15m
Mumbai 20m
Lagos 11m

Copyright © Myriad Editions

23 CULTURAL LOSSES

Arctic
Indigenous people living in settlements such as this will find it increasingly difficult to maintain traditional hunting and fishing skills if the sea-ice melts, affecting seal and polar bear populations. A northward shift in vegetation zones will also take with it their other traditional food source: the tundra-grazing caribou and reindeer.

Climate change is threatening sites that represent the world's cultural and historical heritage, and may also relegate to history some climate-sensitive activities that once reinforced shared values and memories.

There is no comprehensive inventory of important cultural and historical sites under threat from climate change. Many are located in coastal areas, where the threat of sea-level rise and retreating coastlines suggests, at the very least, a new set of conservation challenges. The potential for more frequent floods and more intense storms poses greater threats to cherished buildings, monuments, archaeological sites, and other material traces of history and heritage. Elsewhere, changes in temperature and humidity, as well as threats of extreme weather events, pose conservation challenges.

Many native cultures must work and struggle to maintain traditions that may be threatened by climate change. In the Arctic, potential shifts in species, reduction in ice coverage, and changes to seasons pose risks to shared practice and traditions. Some communities, forced to relocate because of coastal erosion, are facing being separated from family and friends, having to travel further to work, and other, perhaps small, things that affect quality of life. At another scale of disruption, within small island states, saltwater intrusion, increased erosion, and decreased crop productivity make relocating citizens to another nation an increasingly significant possibility. The idea of a place called home is so important that to think of letting go is nearly unimaginable.

Other possible impacts do not reach to the core of cultures in the same way, but they affect events and activities that hold meaning for many. These may involve favorite foods, such as maple syrup; the setting for a holiday, such as a white Christmas; or sporting activities, such as ice-skating. It may be that the change in seasonal climate that signals the time for a holiday comes before that scheduled on the calendar, or nostalgic remembrance of summer nights will not be replicated. Thus, these links to our histories may diminish.

Such impacts are rarely represented in economic estimates of the cost of climate change. Putting a dollar value on these potential losses is both controversial and impossible.

Thoreau's Woods

The flora of the area around Walden Pond, made famous by philosopher and naturalist Henry David Thoreau, appears to be changing due to climate change.

Chan Chan, Peru

The capital of the pre-Hispanic Chimu Kingdom is one of the most important earthen architectural sites, but the highly decorated earthen walls of palaces, temples, and other structures are at risk from increased rainfall and humidity.

Northeast USA

Temperature and precipitation changes threaten sugar maples and maple-syrup production, as well as winter sports and recreation such as skiing and ice fishing.

Scotland, UK

A survey found 12,000 sites vulnerable to coastal erosion, which is accelerated by sea-level rise. Archaeologists expect that 600 sites will be of exceptional importance.

Netherlands

The Eleven City Tour is a 200-kilometer (125-mile) ice-skating marathon along a circuit of canals connecting 11 towns in the northern province of Friesland. It can only be raced when ice conditions are sufficient. That has only occurred only 15 times since 1909, and not since 1997.

Mt. Everest

When Edmund Hillary returned to Everest in 2003 he remarked that, whereas in 1953 snow and ice had reached all the way to base camp, it now ended five miles away. Local people are worried that if Everest loses its natural beauty, tourists will stay away, destroying local livelihoods.

Czech Republic

In 2002, flooding across Europe damaged concert halls, theatres, museums, and libraries. An estimated half a million books plus, in addition to archival documents, were damaged. Climate change may bring more flooding and further losses to some areas.

Cherry Blossom, Japan

Cherry festivals have been an important spring event in Japan for over 1,000 years, but the timing of the event depends on when the trees blossom. Outside Tokyo cherry trees flowered an average 5.5 days earlier in 2005 than in 1981.

Alexandria, Egypt

The monuments of Alexandria, including the 15th-century Qait Bey Citadel, are threatened by coastal erosion and the inundation of the Nile delta region, caused by sea-level rise.

Tuvalu

Forced relocation by sea-level rise is raising questions about nationality, identity and culture as the population faces the threat of statelessness.

Venice, Italy

The structural integrity of many buildings is being damaged by frequent flooding. Current flooding is caused mostly by land subsidence, but rising sea levels exacerbate the problem. Engineering efforts are underway to protect the city.

West Coast National Park, South Africa

The oldest human footprints, estimated to have been made 117,000 years ago, were found here and removed for safe keeping, but other, as yet undiscovered, archaeological evidence is at risk from sea-level rise.

Thailand

Severe flooding currently threatens historical sites. In northeastern Thailand, floods have damaged the 600-year-old ruins of Sukothai, the country's first capital, and the ruins of Ayutthaya, which served as the capital from the 14th to 18th centuries.

PART 5 Responding to Change

The challenges of responding to climate change are unprecedented, but not insurmountable. Key technologies, finance, institutions, and leadership are coming forward and provide greater hope for more effective solutions than in the past. Equally, there is much to be done, and delay will result in higher future costs.

There are two goals in terms of avoiding dangerous levels of climate change: to reduce greenhouse-gas emissions and to ameliorate the impacts of changes in climatic resources and disasters. These goals – mitigation and adaptation – are not substitutes. Both are urgent, and both are required in order to provide an effective response to climate change.

Adaptation is moving forward on all fronts. Countries are preparing national strategies and including adaptation as a major priority in forward plans. Local communities are active in reducing current disaster risks, while ecosystem-based adaptation accelerates action on sound natural resource management; both are beginning to anticipate future changes. Yet, 2010 was one of the mostly costly years ever in terms of the impact of weather-related disasters. Uncertainty regarding the future is the reason for action now: relying on strategies developed for past climate conditions is not adequate.

Technical options for reducing greenhouse-gas emissions are available, and many more are under development. Economic growth is not inevitably tied to increases in emissions. The greater diversity of renewable energy technologies in more recent years opens more opportunities for low-carbon development. For developing countries, investing early in a less carbon-dependent infrastructure offers the potential for long-term savings, and for immediate benefits from reduced air pollution and less dependency on energy imports.

The world's deep dependency on oil is manifest in the increasing prices and extreme efforts to obtain it. A range of new technologies are extracting oil, whether with greater risks from deep-sea drilling or significant environmental costs from hydraulic fracturing of oil shale. Decades of investment in road transportation and private vehicles has created a path of dependence that will be expensive to alter. Stabilizing climate change will require strong policies to set limits to emissions and promote vastly increased investments in low-carbon technology. Not all technological adjustments will be easy.

Large-scale responses such as seeding the atmosphere to increase reflection of sunlight, called geo-engineering, have been reviewed in the past few years. Such approaches are still in their infancy in terms of proven feasibility and have many social, economic, and technical issues before they could be widely adopted. Hopefully, other effective responses will be the first choice before relying on unproven interference in the Earth's climate.

Events in the past few years have been mixed. Many companies have demonstrated the cost effectiveness of reducing energy use, switching fuels, and controlling greenhouse-gas emissions. And climate-response companies, partnerships, services, and initiatives are growing rapidly. Low-carbon technologies are a growth area for investment, with a sense of a shift away from nuclear power generation and risky oil extraction. On the other hand, the pace of international commitments and actual reductions in emissions and climate risks is not encouraging.

> We are all part of the problem of global warming. Let us all be part of the solution.

Ban Ki-Moon
Secretary General of United Nations, April 13, 2008

Weather-related disasters cause on average more than

100,000

deaths and over

$100 billion

in economic losses per year

The impacts of variations in our climate have serious consequences for people's lives and livelihoods. As climate change becomes more marked, the need for adaptation will be ever greater, and more difficult.

In the short-term, the most serious climate impacts are disasters, which increase people's vulnerability, making it even more difficult for them to adapt to future climate change. Current climatic disasters are estimated to cause population displacement of some 20 million people each year. Although the number of people affected continues to rise, disaster risk-reduction measures have ensured that fewer people die, due to early warning systems, evacuation plans and shelters.

The most vulnerable people are those beset by poverty who rely on agriculture, water, and natural resources for their livelihoods. Community Based Adaptation supports sound development while simultaneously reducing the impact of climate change. For instance, given sufficient warning of a coming dry season, farmers have the opportunity to plant drought-resistant crops.

Longer-term changes in resources are more permanent, although less deadly. Managing natural resources wisely is part of an Ecosystem Based Adaptation strategy. Protecting steep slopes from erosion results in less siltation of rivers and reservoirs and more clean water for everyone.

Sectoral strategies – such as those for agriculture, coastal zones, energy, health, urban infrastructure, and water – require planning at all levels. Migration due to melting of permafrost, coastal erosion, and sea-level rise may be unavoidable. However, planning ahead can reduce the costs of replacing infrastructure such as roads and pylons, restrict major development to less dangerous zones, and diversify economic activities. Planners need to balance the costs and benefits, especially between "hard" investments, such as major engineering works, and "soft" responses that build in flexibility to manage a wide range of possible impacts.

Adaptation also takes advantage of new climates. For instance, farmers in northern Europe are enjoying the benefits of longer growing seasons, especially for horticulture.

> There is a very human tendency to wish away such dire prognostications and even to question the underlying science. But the science is now quite firm. People need to be told how it will affect them in their country…

Nitin Desai
Member of the Prime Minister's Council on Climate Change, India

Hurricane Frances
Hurricane Gustav
Hurricane Ike
Floods
Hurricane Jeanne
Hurricane Charley
Hurricane Rita
Hurricane Katrina
Floods
Hurricane Ivan
Hurricane Stan
Hurricane Wilma

Lost ecosystems increase climate impacts

Where mangroves and coastal ecosystems have been destroyed, homes, hotels, and urban infrastructure, such as the seafront in the Bocagrande area of Cartagena, Colombia, are increasingly vulnerable to sea-level rise and coastal storms. Sea-walls and erosion-control measures are costly and often not effective.

CLIMATIC DISASTERS
Number of weather-related disasters
2000–10

	Number of weather-related disasters
■	100 or more
■	50 – 99
■	10 – 49
■	fewer than 10
	none or no data

Disasters 2001–10 involving:

$ damage of $5bn or more

◗ 1,000 or more people dead or missing

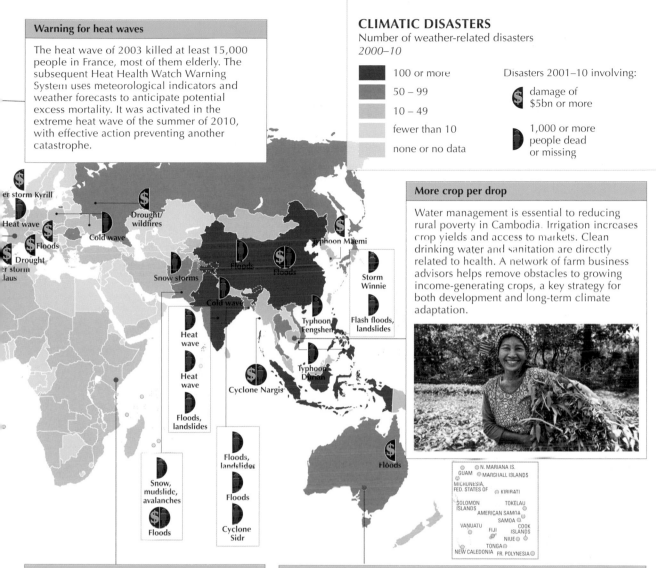

Heat wave

Floods

Drought

er storm Kyrill

er storm laus

Cold wave

Drought/ wildfires

Snow storms

Cold wave

Heat wave

Heat wave

Floods, landslides

Cyclone Nargis

Snow, mudslide, avalanches

Floods

Floods, landslides

Floods

Cyclone Sidr

Typhoon Maemi

Floods

Storm Winnie

Flash floods, landslides

Typhoon Fengshen

Typhoon Durian

Floods

More crop per drop

Water management is essential to reducing rural poverty in Cambodia. Irrigation increases crop yields and access to markets. Clean drinking water and sanitation are directly related to health. A network of farm business advisors helps remove obstacles to growing income-generating crops, a key strategy for both development and long-term climate adaptation.

N. MARIANA IS.
GUAM MARSHALL ISLANDS
MICRONESIA, FED. STATES OF KIRIBATI
SOLOMON ISLANDS TOKELAU
AMERICAN SAMOA
SAMOA
VANUATU COOK ISLANDS
FIJI NIUE
TONGA
NEW CALEDONIA FR. POLYNESIA

Women's groups manage natural resources

A strong tradition of supporting women's groups to manage water, trees and soil erosion in Kenya offers lessons for adapting to future climate change. Capturing rainwater from roofs alleviates sporadic drought and improves health.

Tackling the Big Dry

From 2002 to 2009, a catastrophic drought gripped Australia, particularly in the Murray Darling River Basin in Australia. National and regional responses have included new legislation and regulatory reform; water pricing, conservation and rationing measures; and new infrastructure. Public awareness of the drought, in part as an icon of climate change impacts, has been global. Water consumption dropped dramatically, although long-term prospects are uncertain.

And then the floods hit in 2011, linked to El Niño/Southern Oscillation – a fluctuation in the ocean temperatures associated with droughts and floods. Better warnings would have helped – for both the drought and floods.

25 BUILDING CAPACITY TO ADAPT

Learn, share, connect worldwide

The British Council works to develop International Climate Champions who raise awareness and motivate action around the world. These young people attended the International Youth Forum on Climate Finance in Shanghai in 2010, and drew up proposals for addressing ways of reducing emissions, which they shared with delegates at COP-16 later that year.

Climate champions in front of the UK pavilion, Shanghai Expo 2010.

As individuals and institutions grapple with the challenge of adapting to climate change, they need the capacity to develop effective strategies and actions. Plans are being implemented at all levels, from local, to national and regional, and the wide range of people and institutions involved require targeted support.

Global networks on adaptation are expanding and building capacity, taking many forms. The United Nations leads several efforts, related to the Nairobi Work Programme on Impacts, Vulnerability and Adaptation, organized by the secretariat to the UN Framework Convention on Climate Change (UNFCCC). The UN Development Programme and UN Environment Programme cooperate in developing countries through the Global Adaptation Network, Adaptation Learning Mechanism, and in-country projects. The Global Environment Facility at the World Bank has established a Pilot Programme on Climate Resilience to scale up adaptation efforts. Many development agencies have global and regional networks to support research, build capacity, and coordinate lessons learned. For example, the Global Initiative on Community Based Adaptation brings together NGOs to share experience and inform policy relating to development, poverty reduction, and equity issues.

National planning has been reported in National Communications to the UNFCCC. Developed countries are now in their fifth round of reporting. Countries such as the Netherlands and the UK have quite advanced adaptation plans. In contrast, developing countries have only recently begun to consider adaptation. Least-developed countries gained useful experience through National Adaptation Programmes of Action that identified priority sectors and pilot projects. However, it has taken a decade from inception of the NAPA programme to the first projects being implemented. Local government initiatives are instrumental in building capacity.

Rather than expecting to have all the answers, now, a practical approach is "act, learn, then act again", sharing information and successes. New institutions are being established already, including global and regional networks, as well as specialized efforts to improve the knowledge base to support decision making. Building institutional competence takes many years. Climate adaptation cannot be solved with a quick fix.

California

The California Adaptation Strategy developed from an Executive Order by Governor Schwarzenegger in 2008, has reviewed sectoral risks and prioritized actions by state agencies and other stakeholders. California's long history of state climate assessments and the early focus on mitigation provide the groundwork for the Climate Action Team to focus on adaptation. The strategic plan emphasizes the use of existing regulations, coordination across agencies, and improving the knowledge base. CalADAPT helps raise awareness.

Supporting local adaptation

A coffee picker in El Salvador, working on a cooperative certified by the Rainforest Alliance, whose mission is to support responsible land management and sustainable livelihoods in the world's forested areas.

Peru

Climate change impacts in Peru may cost $855 billion by 2050. Several agencies are collaborating on a four-year project, supported by a $33-million budget from the Special Climate Change Fund of Global Environmental Facility, to adapt to the impact on water resources of the rapidly retreating Andes glaciers. Programs in the southern highlands are highlighting the importance of reducing the risk from disaster.

Urban planning

The Asian Cities Climate Change Resilience Network, sponsored by the Rockefeller Foundation, is preparing for the predicted impacts of climate change. By 2012, the network aims to test a range of actions, build a knowledge base of lessons learned, and assist cities to build resilience to climate change.

NATIONAL ADAPTATION PLANNING
Status of reporting to the UNFCCC
as of December 2010

- submitted 5th National Communication
- submitted earlier National Communication
- no National Communication submitted

National Adaptation Programme of Action

- ● NAPA Priority Project approved
- ◖ NAPA Priority Project proposed
- ○ NAPA plan submitted

ICELAND

NORWAY SWEDEN FINLAND
ESTONIA
DENMARK LATVIA
UK RUSSIA LITHUANIA
IRELAND NETH. POLAND BELARUS
GERMANY
BEL. CZ. UKRAINE
LUX. LIECHT. SLK. MOLDOVA
FRANCE AUS. HUN. ROM.
S. MARINO SLO. CRO. B.H. S.
MONACO M. BUL.
ITALY ALB. MAC.
PORTUGAL SPAIN GREECE TURKEY
TUNISIA MALTA CYPRUS SYRIA
LEBANON IRAQ
MOROCCO ISRAEL JORDAN
ALGERIA LIBYA EGYPT KUWAIT
BAHRAIN
QATAR
SAUDI ARABIA UAE
OMAN
MAURITANIA MALI NIGER YEMEN
CAPE ERITREA
VERDE
SENEGAL CHAD SUDAN DJIBOUTI
GAMBIA
GUINEA BURKINA
BISSAU GUINEA FASO NIGERIA
SIERRA LEONE CÔTE BENIN C.A.R. ETHIOPIA
GHANA SOMALIA
D'IVOIRE CAMEROON
LIBERIA TOGO EQUATORIAL UGANDA
GUINEA KENYA
SÃO TOME GABON RWANDA
& PRINCIPE CONGO DEM. REP. BURUNDI SEYCHELLES
OF CONGO
TANZANIA COMOROS
ANGOLA MALAWI
ZAMBIA MAURITIUS
ZIMBABWE MADAGASCAR
NAMIBIA
BOTSWANA MOZAMBIQUE
SWAZILAND
SOUTH
AFRICA LESOTHO

RUSSIA
KAZAKHSTAN
MONGOLIA
GEORGIA UZBEK. KYRGYZSTAN NORTH JAPAN
AZERBAIJAN KOREA
ARMENIA TURKMEN. TAJIKISTAN SOUTH
AFGHANISTAN KOREA
IRAN CHINA
BHUTAN
PAKISTAN NEPAL
BANGLADESH BURMA
INDIA LAOS
THAILAND VIETNAM
CAMBODIA
SRI LANKA PHILIPPINES
MALDIVES BRUNEI
MALAYSIA
SINGAPORE
INDONESIA
EAST TIMOR
PAPUA
NEW
GUINEA
AUSTRALIA
NEW
ZEALAND

MICRONESIA, MARSHALL ISLANDS
FED. STATES OF
NAURU KIRIBATI
SOLOMON TUVALU
ISLANDS
SAMOA
VANUATU FIJI NIUE
TONGA COOK
ISLANDS
PALAU

African Climate Policy Centre

The ACPC is a regional "centre of excellence" designed to promote successful climate policy for both mitigation and adaptation. Established in 2008 by the African Union Commission, UN Economic Commission for Africa, and African Development Bank, it promotes knowledge generation, sharing and networking; advocacy and consensus building; and advisory services and technical cooperation. It will link to existing networks and initiatives, such as Africa-Adapt.net, an online community hosted by ENDA, the University of Cape Town global database on climate change scenarios, and the African Centre for Technology Policy Studies climate change network in over 20 countries.

Bangladesh

Bangladesh has a long tradition of assessment and action on climate-related hazards. Action two decades ago dramatically reduced the toll of tropical cyclones. Recently, the government adopted a 10-year program covering social protection, disaster management, infrastructure, research and knowledge management, low-carbon development, and institutional strengthening. The total cost is expected to be $5 billion over five years.

In many places, local and regional authorities are developing more aggressive emission reduction policies than federal governments are. This includes a commitment to:

- conduct an energy and emissions inventory and forecast
- establish an emissions target
- develop a Local Action Plan
- implement policies and measures to reduce emissions
- monitor and verify the results

BENEFITS

Many strategies to reduce greenhouse gas emissions also benefit local residents by:

- generating financial savings through energy and fuel efficiency
- encouraging economic development and job creation
- preserving green spaces
- reducing air pollution
- decreasing traffic congestion

Cities around the world are taking action. They are not waiting for national governments to debate the next step. They are signing their own commitments to reducing greenhouse gas emissions, and many organizations are proving supportive.

City leaders are becoming increasingly aware of their potential to reduce greenhouse gas emissions and the risks their cities face – heat waves, water supply disruptions, flooding and more severe storms among them. Over 2,800 local and city governments contributed to a catalog of climate-change action commitments in preparation for the UN Climate Convention meeting of parties in 2009. The C40 cities partnership represents some of the world's largest cities. The Covenant of Mayors includes over 1,900 European municipalities committed to 50 percent reduction in greenhouse gas emission by 2020. Worldwide, there are over 1,200 local governments working to implement sustainable development through networks supported by ICLEI – Local Governments for Sustainability.

Half the people in the world now live in cities, which together account for between 30 and 40 percent of global CO_2 emissions. While per capita emissions of urban dwellers are typically less than that of their rural counterparts, large cities represent significant concentrations of people and resource consumption. For example, greenhouse gas emissions from Tokyo are comparable in volume to that of countries such as Denmark and Norway. The world's population is continuing to become more urban, so innovation in cities will be an increasingly important part of mitigation and adaptation activities.

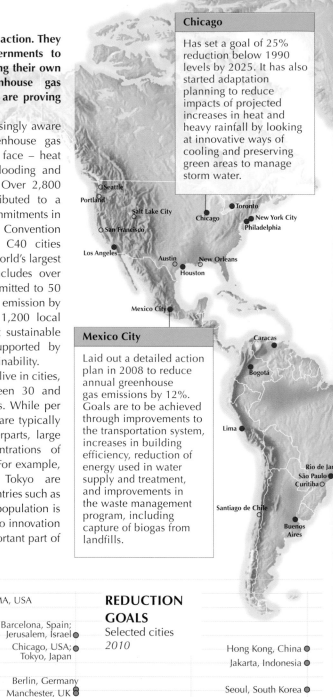

Chicago

Has set a goal of 25% reduction below 1990 levels by 2025. It has also started adaptation planning to reduce impacts of projected increases in heat and heavy rainfall by looking at innovative ways of cooling and preserving green areas to manage storm water.

Mexico City

Laid out a detailed action plan in 2008 to reduce annual greenhouse gas emissions by 12%. Goals are to be achieved through improvements to the transportation system, increases in building efficiency, reduction of energy used in water supply and treatment, and improvements in the waste management program, including capture of biogas from landfills.

REDUCTION GOALS
Selected cities
2010

reduction vs. target years chart:

- Toronto, Canada — 0%, 2012
- Phoenix, AZ; Baltimore, MA, USA — -5%, 2012
- Mexico City, Mexico — -10%, 2012
- Madrid, Spain — -15%, 2012
- Melbourne, Australia — -20%, 2015
- Johannesburg, South Africa — -15%, 2011
- Barcelona, Spain; Jerusalem, Israel — -20%, 2020
- Chicago, USA; Tokyo, Japan — -25%, 2020
- Berlin, Germany — -40%, 2020
- Manchester, UK — -40%, 2020
- Torino, Italy — -50%, 2015
- London, UK — -60%, 2020
- Hong Kong, China — -25%, 2030
- Jakarta, Indonesia — -30%, 2030
- Seoul, South Korea — -40%, 2030
- Vienna, Austria; Oslo, Norway — -50%, 2030

target years: 2011 2012 | 2015 | 2020 | 2025 | 2030

London

Has set the ambitious target of reducing emissions by 60% from 1990 levels by 2025. The Mayor has also proposed an adaptation strategy aimed at increasing understanding of risks to the city, and buffering the effects of floods, hot weather, and drought. The plan gives particular attention to risks to health, the environment, business, and urban infrastructure, such as water and energy supply.

C40 CITIES
2010

- ● C40 large city
- ○ C40 affiliate city

The C40 partnership is designed to share information, best practices, and ideas, as well as provide leadership on tackling climate change. It works with the Clinton Climate Initiative to support efforts to reduce greenhouse gas emission through a range of energy efficiency and clean energy programs.

Stockholm
Copenhagen
Amsterdam Berlin
London Rotterdam Warsaw
Paris Heidelberg
Basel
Milan
Barcelona Rome
Madrid
Athens
Istanbul
Moscow
Cairo
Lagos
Addis Ababa
Johannesburg
Delhi
Karachi
Mumbai
Dhaka
Bangkok
Hanoi
Hong Kong
Ho Chi Minh City
Jakarta
Beijing
Seoul
Changwon
Yokohama
Tokyo
Shanghai
Melbourne
Sydney

Tokyo

In 2010, it initiated mandatory CO_2 reduction for large-scale industrial and commercial establishments, such as office buildings. In 2011, it plans to launch what will be the first cap and trade program in Asia. Other plans include incentives for renewable energy, electric cars, and green space conservation and extension.

Lagos

In 2010, held its second planning summit on climate change, bringing together 1,200 stakeholders to discuss adaptation and mitigation strategies and recommend actions.

Johannesburg

The city is investing in climate-friendly retrofitting of almost 100 city buildings and low-income housing, including solar water heaters (right), as well as developing a greenhouse-gas inventory of the city.

Mumbai

In April 2010, the city government contracted an analysis of projected climate change impact on hydrology and water resources, agriculture, coastal areas, marine ecosystems and livelihoods, including possible migration. The results, due in 2012, will inform adaptation planning with more locally specific information on vulnerability and potential range of impacts.

Sydney

Has become carbon neutral by reducing carbon emissions from its vehicle fleet, increasing efficiency of properties and street lighting, and using greenhouse gas offsets.

GROWTH IN WIND ENERGY

Cumulative installed capacity
2005 & 2009
gigawatts

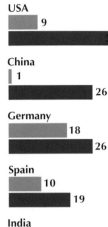

2005 ■   **2009** ■

USA
9
35

China
1
26

Germany
18
26

Spain
10
19

India
4
11

Renewable energy could be the key to economically and socially sustainable societies.

Investment in technologies to convert solar light and heat, and the force of wind and water into usable energy sources is expanding rapidly, more than doubling between 2006 and 2009, from $63 billion to $150 billion. Even so, only about 5 percent of the 8,286 million tonnes of oil equivalent energy used worldwide comes from renewable sources. Although this estimate under-represents small-scale energy systems, it is clear that renewable sources could make a much larger contribution.

The greatest increase in investment generally occurs in the countries, provinces, and states with policies that encourage energy companies to develop renewable energy sources. These include subsidies, tax advantages, and a renewable portfolio standard – a requirement for a minimum proportion of energy to be generated from renewable sources. Feed-in tariffs are also used to encourage consumers to invest in small-scale installations by offering an attractive price for surplus energy contributed to the national grid.

The International Renewable Energy Agency (IRENA), established in January 2009 with a mission to promote renewable energy, had 142 countries and the EU as signatories by the end of its first year.

Energy generated by wind expanded worldwide from less than 10 GW global capacity in 1996 to over 159 GW, increasing by 41 percent in 2009 alone. China doubled its installed capacity for six consecutive years up to 2009.

Photovoltaics, a technology that was under-developed until the 1990s, is contributing to the grid of many countries. Installed capacity quadrupled between 2006 and 2009. These solar cells are flexible enough to provide electricity for rural communities not linked to a large electricity grid. Solar energy is also being used to produce thermal energy to heat water. China, Turkey, Israel, and Japan are major users of this technology, and it is expected to increase in countries with favorable policies.

Over 2 million buildings in 76 countries are using geothermal energy. The increasingly popular biodiesel fuels also offer significant reductions in emissions of greenhouse gas.

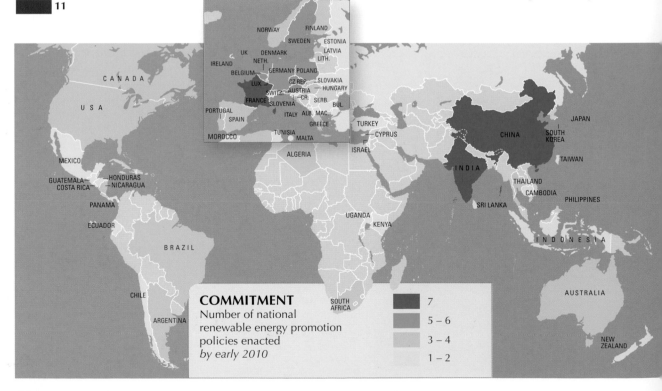

COMMITMENT
Number of national renewable energy promotion policies enacted
by early 2010

■ 7
■ 5 – 6
■ 3 – 4
■ 1 – 2

RENEWABLE ENERGY SOURCES
end 2009

■ electricity produced, GW installed capacity

■ heat produced, GWth installed capacity

Small-scale hydro

Hydro systems generate electricity from running water. They can provide power for isolated villages, or feed power into the electricity grid.

60

Wind

Wind turbines of varying sizes are used to generate electricity, for the national grid or for isolated communities.

159

installed capacity

Biomass

Plant material – purpose grown or waste – can be burned or fermented, and used to generate electricity or heat. The CO_2 released is the same amount as was removed from the atmosphere during the plant's lifetime, so biomass is considered carbon neutral. However, changing the focus of agriculture to fuel production involves tradeoffs with food, forests, and water.

270

54

Geothermal

In geologically active areas, the Earth's intense heat can fuel power plants. Elsewhere, Earth temperature, which remains relatively constant at depths greater than 10 meters below the surface, can be used to heat and cool buildings.

11 **60**

Solar

Thermal panels convert the sun's radiation into heat, often warming water.

Photovoltaic panels (below) convert the sun's radiation into electricity.

180

21

installed capacity

Tide, wave, ocean

The movement of the sea can be used to generate electricity. **0.3**

RENEWABLE ELECTRICITY PRODUCTION

Share of production by greenhouse gas neutral technologies for selected countries
end 2009
gigawatts

■ geothermal
■ solar photovoltaic grid
■ wind

■ biomass
■ small-scale hydro

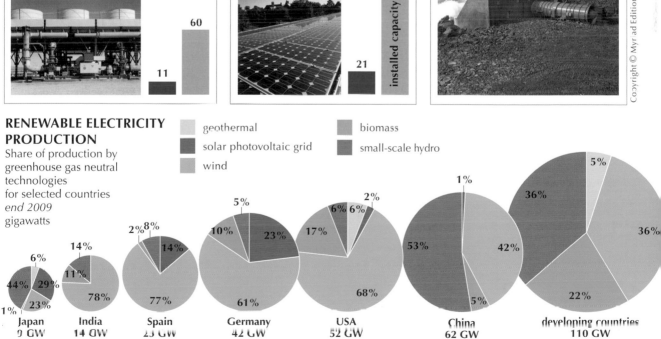

Japan
9 GW
6% · 29% · 44% · 1% · 23%

India
14 GW
14% · 11% · 78%

Spain
23 GW
2% · 8% · 14% · 77%

Germany
42 GW
5% · 10% · 23% · 61%

USA
52 GW
17% · 6% · 6% · 2% · 68%

China
62 GW
1% · 53% · 42% · 5%

developing countries
110 GW
5% · 36% · 36% · 22%

CLIMATE PRINCIPLES

CLIMATE PRINCIPLES

The Climate Group principles commit organizations and businesses to low-carbon futures:

1 To have a robust low-carbon strategy or position, and to manage operational carbon emissions.

2 To develop commercially viable approaches to ensure climate and carbon issues are addressed.

3 To engage others to support the growth of a low-carbon economy, where consistent with corporate policies on public engagement.

Developing countries need to overcome poverty and achieve a level of economic growth, but this will increase emissions unless it can be achieved more efficiently than the historical paths taken by developed countries.

A country's "carbon intensity" is a measure of the efficiency of its economic output with respect to its carbon dioxide (CO_2) emissions. Some countries release relatively little carbon dioxide into the atmosphere when compared with their economic output. They have a low "carbon intensity" and their economies are considered to be comparatively clean.

Industrialization has tended initially to develop through industries with high carbon dioxide emissions, such as shipping, steel and manufacturing. Only as an economy matures, with the growth of hi-tech industries, and the use of more efficient technology to process natural resources, does high economic output become associated with less pollution.

This historic pathway need not be taken by newly industrializing countries, however. A growing awareness about greenhouse gases, and the implementation of policies to force corporations to be environmentally responsive are essential. In many cases, future emissions can be dramatically reduced in developing countries through a mix of cost-effective strategies.

Developing countries with large and rapidly growing economies are pivotal. Brazil, China, India, Indonesia, Mexico, Russia, and South Africa accounted for 36 percent of CO_2 emissions in 2008, and carbon intensities in China, India, and Indonesia were five to eight times that of the UK.

More efficient technology needs to be introduced. The Clean Development Mechanism, part of the Kyoto Protocol, and various market instruments have spawned a dramatic increase in interest in technology, but negotiations to ensure that efficient technologies are widely available in emerging and developing economies have been stymied by global barriers of intellectual property rights and proprietary pricing.

POTENTIAL REDUCTIONS
In annual emissions of selected industrializing countries under two scenarios

	Potential reduction in annual emissions million tonnes CO_2e		Possible reduction by 2030 from use of available technologies	Plausible reduction by 2030 with carbon prices at $20 a tonne	Percentage contribution of key sectors to potential reductions
	2005	2030			
China	6,800	14,500	46%	32%	power generation 42% industry 24%
India	1,570	5,740	45%	31%	power generation 36% industry 26%
Brazil	1,260	1,700	52%	41%	forests >50%, power 18%
Indonesia	2,250	3,590	63%	55%	forests/peat 70%
Mexico	660	1,100	40%–50%	34%–45%	power generation 20–25% transport 15–25%
South Africa	450	1,100	55%	>40%	range across sectors (particularly power sector)
Russia	2,150	2,990	51%	33%	power generation 20% buildings 21%

◀◀ 48–49 Fossil Fuels; 52–53 Transport

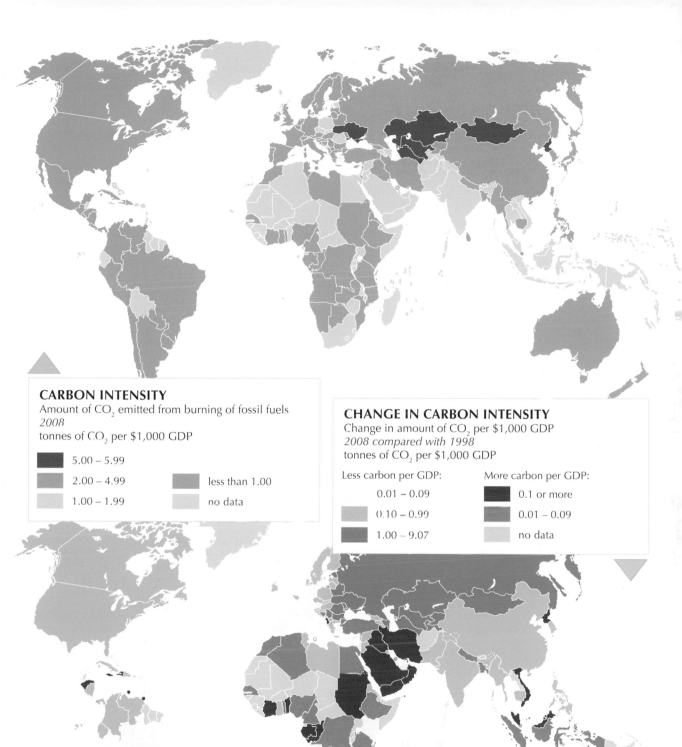

CARBON INTENSITY
Amount of CO_2 emitted from burning of fossil fuels
2008
tonnes of CO_2 per $1,000 GDP

5.00 – 5.99	
2.00 – 4.99	less than 1.00
1.00 – 1.99	no data

CHANGE IN CARBON INTENSITY
Change in amount of CO_2 per $1,000 GDP
2008 compared with 1998
tonnes of CO_2 per $1,000 GDP

Less carbon per GDP:
- 0.01 – 0.09
- 0.10 – 0.99
- 1.00 – 9.07

More carbon per GDP:
- 0.1 or more
- 0.01 – 0.09
- no data

COUNTING CARBON IN FORESTS

Calculating the carbon content for a single tree gives some idea of the complexity of estimating how much carbon is locked up in forests around the world.

A tree's height and diameter need to be measured in order to calculate the volume. An estimate of its carbon content can then be made by reference to tables listing the weight/unit volume and carbon/unit weight for each species. To account for future carbon sequestration, the tree's growth rate also needs to be estimated.

Calculating the carbon sequestration potential of the world's 4 billion hectares of forest involves taking into account the variety of tree species, their ages, and their growth rates. In addition, carbon stored in woody debris on the forest floor, and in tree roots and soil need to be factored in. All kinds of assumptions need to be made, but the more detailed and accurate the calculations, the more costly they become.

The allocation of responsibility for reducing emissions, and the crediting of efforts to reduce and store carbon, involve detailed assessment processes. These can be based on the amount of greenhouse gases produced, or the amount consumed, but this simple distinction masks a wide array of technical, sectoral, and country-specific variables, all of which need to be considered in the interests of improving the accuracy of the accounting.

The UN Framework Convention on Climate Change (UNFCCC) regional and national cap and trade systems, as well as some voluntary reporting systems, are designed to collect information on the amount of carbon and other greenhouse gases produced, or the amounts not produced as a result of mitigation efforts. The complex IPCC reporting guidelines fill five volumes, and include details relevant to different sectors (such as energy, industrial processes, agriculture, forestry, other land uses, and waste), categories within sectors (such as transportation), and sub-categories (such as cars). The results are aggregated in national reports to the Secretariat of the UNFCCC, and incorporate reports under the cap and trade system. It is important that all parties have confidence in the design, implementation, and verification of this accounting.

One important example of a mitigation program that requires systematic counting is the "reduction of emissions from deforestation and forest degradation" or REDD+, which aims to increase the role of forest conservation and sustainable management in developing countries. This will involve calculating and monitoring changes in forests, but even establishing the baseline status of forests types involves many technical challenges.

Another area of carbon accounting focuses on the increasing number of emissions trading programs, which are leading to a growing need for refined methods by which corporations can accurately assess their emissions. There are several sources of guidelines, from individual reporting registries, to the IPCC guidance on national reporting. In 2010, the US Environmental Protection Agency also began actively regulating CO_2 as a pollutant under the US Clear Air Act, requiring large emitters across a variety of industries to report emissions of greenhouse gases.

REDD+

REDD+ aims to create financial incentives for developing countries to develop sustainable management systems for their existing forests by establishing an economic value for the carbon stored in the wood and soil. One of the main issues being addressed by the organizations involved is developing (and gaining approval for) the accounting techniques and methodologies needed to verify a project's effectiveness.

Some methods rely on data collected from satellite images, which can measure the extent of forest and land-cover change. Although these images cover vast areas, consistency in the spatial resolution of the images and in the interpretation of the data may be an issue.

Other methods trade off simplifying assumptions, broad coverage, and low cost with more detailed, time-consuming on-the-ground surveys. These involve making inventories of tree species, measuring the density of cover, and taking soil samples. They have the advantage of involving local people, thereby increasing support for the project. Here, a villager nails a number tag to a tree in Monks Community Forest, Cambodia.

Estimating the carbon emissions saved by sustainable management is also difficult, and may have to be done by monitoring forest loss from an equivalent unmanaged area. Another issue to be taken into account is whether the process of deforestation is displaced from the managed area into surrounding areas.

Mosaic of deforested area in Bolivia. Healthy vegetation is shown in red.

CONSUMPTION-BASED ACCOUNTING

The question of whether the producer or the consumer of traded goods is responsible for reducing carbon emissions is a focus of international debate. For example, if a microwave oven is manufactured in China and sold to a family in Sweden, should the emissions associated with its manufacture and transport be added to the carbon account of those who profit from it, or of those who use it? Consumption-based accounting specifies the carbon emissions related to imported and exported products in calculations of a nation's carbon footprint. It also attempts to monitor international "carbon leakage" – increases in emissions in one country driven by emission reduction efforts in another that displace carbon intensive activities, such as manufacturing.

Some countries rely very heavily on imports, which can represent a substantial percentage of their consumption-based emissions. An effort at modeling carbon embodied in international trade found that in 2004, Austria, France, Sweden, Switzerland, and the UK imported more than 30 percent of their consumption-based emissions – over 4 tonnes of CO_2 per person. The USA, then the largest emitter of CO_2 from within its own borders, imported goods embodying emissions that represented an additional 10.8 percent. In China 22.5 percent of its CO_2 emissions were exported, for consumption in other countries. Allocating the responsibility for emissions to those who consume the food or use the products would substantially reduce the national emissions of countries such as China, and increase those of Japan, the USA, a any European countries.

Around **23%** *of CO_2 emissions from fossil-fuel burning in 2004 was released in the production of exported goods*

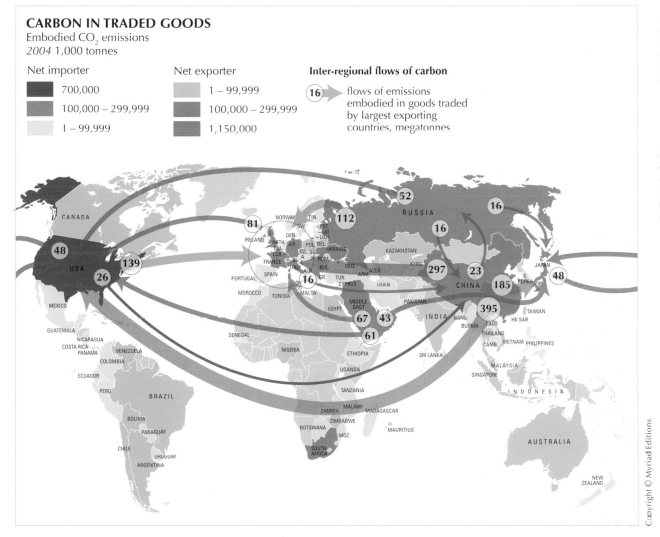

CARBON IN TRADED GOODS
Embodied CO_2 emissions
2004 1,000 tonnes

Net importer
- 700,000
- 100,000 – 299,999
- 1 – 99,999

Net exporter
- 1 – 99,999
- 100,000 – 299,999
- 1,150,000

Inter-regional flows of carbon
16 → flows of emissions embodied in goods traded by largest exporting countries, megatonnes

PART 6 International Policy & Action

Reduction in greenhouse gas emissions to safe levels requires truly international cooperation. Negotiations over climate-change responses recognize complex patterns of causes, relative contributions to the problem, different abilities to contribute to solutions, potential for benefits and losses, and irreversible changes. Laid over existing political differences, global economies and trade, and other factors of international relations, agreements such as the Kyoto Protocol do not progress quickly.

In the past few years, an international consensus has been achieved to limit climate change to 2°C, as signalled in the Copenhagen Accord. However, some believe 2°C still exposes the world to major impacts and unacceptable risk and are pushing for a global target of no more than 1.5°C. In contrast, voluntary commitments as of mid-2011 would result in global warming in the range of 2.5°C to 4°C. The gap is worrying.

While the Kyoto Protocol failed to deliver a lasting solution, there has been considerable progress in building the infrastructure for responding to climate change: the administrative procedures for financing action, monitoring outcomes, reporting progress, and governing carbon markets. These elements and processes are necessary to support cooperation. Awareness of the need for action continues to build.

Most nations have signed international agreements, but they are also faced with forging agreements at home. Meeting reduction targets requires the involvement of local government, small businesses, corporations, and civil organizations; even religious organizations are taking a strong stand to reduce emissions. In many countries, municipalities and companies have organized to reduce emissions and are challenging their national leaders.

Carbon trading is growing, but does not reduce emissions on its own. The growing role of reducing emissions from deforestation and degradation (called REDD+) will change many local to global linkages. International aid provides funding to support some efforts, but it is a very small fraction of development assistance. And finance that takes five years from commitment to implementation can hardly be called "fast-track".

Supporting the capacity of local populations to cope with the challenges of current climatic hazards and longer-term climate change remains a priority. Linking energy efficiency and the management of climatic risks with sustainable development, through education, health care, employment, and information, is essential.

The seemingly endless international road maps should not be confused with effective action. That many organizations and social entrepreneurs are seeking change is a clear sign of hope.

> ...different countries are in different political moments and that is understandable. But on the whole, what is not doable is to evade the responsibility of addressing climate change as a collection of countries, as a community of nations.
>
> **Christiana Figueres**
> Executive Secretary,
> UNFCCC

> The ultimate objective of this Convention ... is to achieve stabilization of greenhouse gas concentrations in the atmosphere at a level that would prevent dangerous anthropogenic interference with the climate system...within a time-frame sufficient to allow ecosystems to adapt naturally..., to ensure that food production is not threatened and to enable economic development to proceed in a sustainable manner.
>
> **UNFCCC**
> Article 2

Most countries have signed the UN Framework Convention on Climate Change (UNFCCC), and have agreed to negotiate effective protocols.

Climate change continues to be high on the international agenda, but there is still much disagreement as to what to do, when, and by whom. Conflicts over relative responsibilities for reducing emissions and funding adaptation continue to slow down negotiations.

The UNFCCC was agreed in Rio de Janeiro in June 1992, and came into force in March 1994. The Convention places the initial onus on the industrialized nations and 12 economies in transition to reduce their own emissions, and to finance developing countries' search for strategies to limit their emissions in ways that will not hinder their economic progress.

The Convention is a flexible framework, clearly recognizing that there is a problem. The first addition to the treaty, the Kyoto Protocol, set targets for reductions in emissions. Adopted in 1997, it came into force in February 2005. The USA ratified the Convention but not the Protocol, whereas Australia finally ratified the Protocol in 2007.

Negotiations occur throughout the year, culminating in the Conference of Parties (COP) and Meeting of Parties of the Kyoto Protocol (MOP) at the end of the year. The Copenhagen meeting in 2009 produced an accord that signals intent to continue negotiations. Countries adopted a voluntary target of limiting global climate change to 2°C. The meeting in Cancún in 2010 made further progress on procedures and finance. However, the binding commitments of the Kyoto Protocol expire in 2012 and there is considerable uncertainty around the next steps.

OBSERVERS

Organizations that are not Parties may observe proceedings (but not closed sessions). The application for observer status lists nine categories of organization, listed here with examples of each:

Business and industry NGOs (BINGOs): American Petroleum Institute, Global Adaptation Institute, Creating Shared Value

Environmental NGOs (ENGOs): Climate Action Network, Greenpeace, International Institute for Environment and Development

Indigenous peoples organizations (IPOs): Aleut International Association, Indigenous Peoples of Africa Co-ordinating Committee

Local government and municipal authorities (LGMAs): Local Governments for Sustainability, Local Government Association

Research and independent NGOs (RINGOs): New Economics Foundation, International Centre for Climate Change Adaptation and Development, Climate Adaptation Knowledge Exchange (CAKE)

Trade union NGOs (TUNGOs): International Trade Union Confederation

Farmers NGOs: Arid Lands Information Network, International Development Enterprises Cambodia

Women and gender NGOs: Women and Gender Constituency in the UNFCCC, Commission on the Status of Women

Youth NGOs (YOUNGOs): International Youth Climate Movement, International Climate Champions

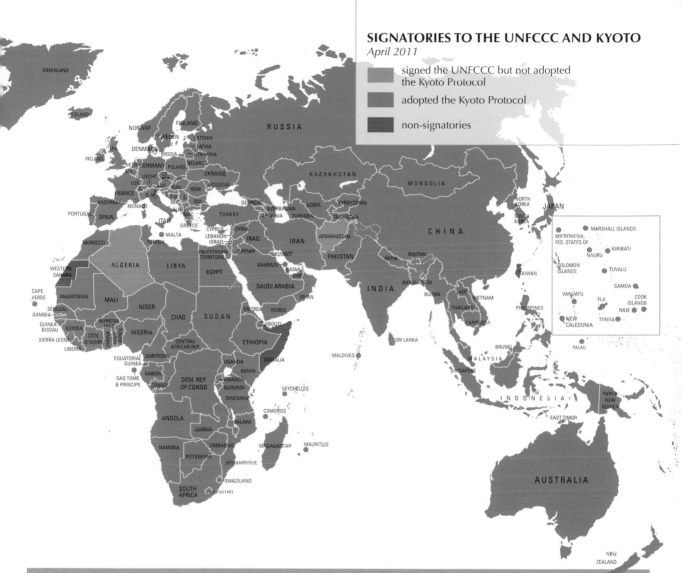

SIGNATORIES TO THE UNFCCC AND KYOTO
April 2011

- signed the UNFCCC but not adopted the Kyoto Protocol
- adopted the Kyoto Protocol
- non-signatories

KEY GROUPINGS WITHIN THE NEGOTIATIONS

Negotiations take place within the UNFCCC framework to develop policies and operational procedures. Countries tend to affiliate themselves to others with similar agendas.

Alliance of Small Island States (AOSIS): a coalition of Small Island Developing States (SIDS): low-lying and small island countries.
BASIC (Brazil, South Africa, India, and China): countries that are a major source of greenhouse gas emissions because of their large populations and economies, but which have lower emissions per capita than developed countries.
Economies in Transition (EIT): includes Russia, the Baltic States, and several states in central and eastern Europe.
European Union: negotiates on behalf of its 27 members, which are also Parties in their own right.
Group of 77: over 130 developing country members (including China), with diverse interests.
Least Developed Countries (LDCs): 48 countries with low incomes, poor human capital and economic vulnerability, especially vulnerable to the adverse impacts of climate change.
Umbrella Group: a loose coalition of non-EU developed countries, critical of the Kyoto Protocol, has included Australia, Canada, Iceland, Japan, New Zealand, Norway, Russia, Ukraine, and the USA.

Various other groups are active at times: the **Environmental Integrity Group** (EIG) of Mexico, South Korea, and Switzerland sought to strengthen global targets. The **Organization of Petroleum Exporting Countries** (OPEC) seeks to limit the cost of mitigation.
The **African Union** endorsed a climate change plan in 2007, and was increasingly effective in promoting an Africa-wide position.
The **Coalition of Rainforest Nations** seeks to facilitate consensus on issues related to domestic and international frameworks for rainforest management, biodiversity conservation and climate stability.

31 MEETING KYOTO TARGETS

INCREASED EMISSIONS

Share of annual emissions*
1990 & 2005
million tonnes CO$_2$e

* including from land-use change

■ Annex I countries

□ non-Annex I countries

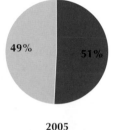

**1990
total 35,301**

49% 51%

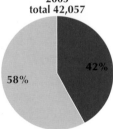

**2005
total 42,057**

58% 42%

Not quite half of Annex I countries have met or are close to meeting their Kyoto commitments, and even the agreed targets fall far short of stabilizing greenhouse gas emissions at levels considered to be safe.

Under the Protocol, the industrialized countries, and those whose economies were considered to be in transition in 1997, the so-called Annex I countries, made commitments to greenhouse gas reductions. Some are making progress towards their goals. Many of the economies making the transition from communist to a market economy continue to experience steep declines in emissions as their economies struggle. Other countries have actually increased their emissions, some quite dramatically.

Building on the United Nations Framework Convention on Climate Change (UNFCCC) principle of "common but differentiated responsibilities," these 40 countries were to lead the way in reduction efforts because they have been emitting greenhouse gases into the atmosphere for a long period of time, and are considered able to afford reductions. The

signatory countries are committed to reducing their greenhouse gas emissions to a combined average of 4.7 percent below their 1990 levels by a target date between 2008 and 2012. The USA, Turkey, and Belarus are Annex I countries, but they are not participating. The refusal of the USA to ratify the Protocol means that about 30 percent of current emissions by Annex I countries are not covered under this agreement.

The uneven progress leaves many issues of equity and commitment among countries unresolved. In the meantime, contributions from non-Annex I countries now represent more than half of the annual total emissions.

The reductions under the Kyoto Protocol were viewed only as a first step because reductions of between 60 percent and 80 percent by 2050 will be needed to avert more serious climate-change impacts. The Kyoto Protocol commitments are set to expire in 2012. As the international community debates and struggles to draft the successor agreement, a great deal of attention has been directed to evaluating the lessons and progress of these commitments.

LONG-TERM EFFECT OF GREENHOUSE GAS EMISSIONS
Assuming a rapid decline after 2050

Predicted impacts
Even if CO$_2$ emissions are rapidly reduced after 2050, and CO$_2$ concentrations in the atmosphere are stabilized within 100 years, there are long-term consequences for global temperature and sea-level rise.

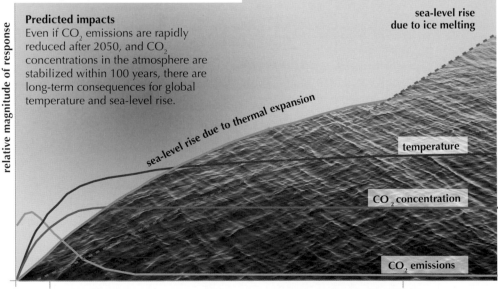

relative magnitude of response

sea-level rise due to ice melting

sea-level rise due to thermal expansion

temperature

CO$_2$ concentration

CO$_2$ emissions

2006   2100 *projected*   3100 *projected*

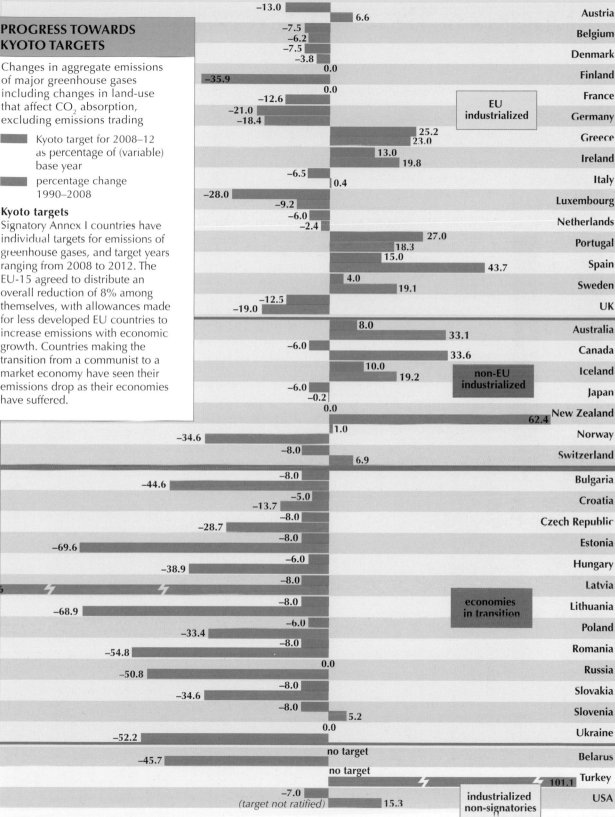

PROGRESS TOWARDS KYOTO TARGETS

Changes in aggregate emissions of major greenhouse gases including changes in land-use that affect CO_2 absorption, excluding emissions trading

▬ Kyoto target for 2008–12 as percentage of (variable) base year

▬ percentage change 1990–2008

Kyoto targets

Signatory Annex I countries have individual targets for emissions of greenhouse gases, and target years ranging from 2008 to 2012. The EU-15 agreed to distribute an overall reduction of 8% among themselves, with allowances made for less developed EU countries to increase emissions with economic growth. Countries making the transition from a communist to a market economy have seen their emissions drop as their economies have suffered.

Country	Kyoto target	% change 1990–2008
Austria	–13.0	6.6
Belgium	–7.5	–6.2
Denmark	–7.5	–3.8
Finland	0.0	–35.9
France	0.0	–12.6
Germany	–21.0	–18.4
Greece	25.2	23.0
Ireland	13.0	19.8
Italy	–6.5	0.4
Luxembourg	–28.0	–9.2
Netherlands	–6.0	–2.4
Portugal	27.0	18.3
Spain	15.0	43.7
Sweden	4.0	19.1
UK	–12.5	–19.0

EU industrialized

Country	Kyoto target	% change 1990–2008
Australia	8.0	33.1
Canada	–6.0	33.6
Iceland	10.0	19.2
Japan	–6.0	–0.2
New Zealand	0.0	62.4
Norway	1.0	–34.6
Switzerland	–8.0	6.9

non-EU industrialized

Country	Kyoto target	% change 1990–2008
Bulgaria	–8.0	–44.6
Croatia	–5.0	–13.7
Czech Republic	–8.0	–28.7
Estonia	–8.0	–69.6
Hungary	–6.0	–38.9
Latvia	–8.0	–310.6
Lithuania	–8.0	–68.9
Poland	–6.0	–33.4
Romania	–8.0	–54.8
Russia	0.0	–50.8
Slovakia	–8.0	–34.6
Slovenia	–8.0	5.2
Ukraine	0.0	–52.2

economies in transition

Country	Kyoto target	% change 1990–2008
Belarus	no target	–45.7
Turkey	no target	101.1
USA	–7.0 (target not ratified)	15.3

industrialized non-signatories

32 LOOKING BEYOND KYOTO

IS IT ENOUGH?
Estimate of total annual emissions in 2020 gigatonnes CO_2e

Emissions total if current national commitments are met **48–55**

Estimated to give a 50:50 chance of holding global temperature increase to 2°C **44–46**

Estimates for the total reductions represented by various commitments are difficult to calculate because they are expressed in many ways, but it would appear that these voluntary agreements are insufficient to meet the goal of restricting the global temperature increase to below 2°C.

The commitments to reduce greenhouse gases agreed in the Kyoto Protocol expire at the end of 2012. Politicians, policy analysts, lawyers, and businesses are now contemplating what international climate change response will look like in a post-Kyoto world.

There were high hopes that the Conference of Parties (COP 15) held in Copenhagen in December 2009 would resolve key issues needed to establish the next protocol for continued and improved international response. Despite growing awareness of the urgency of reducing greenhouse gas emissions, many issues remained unresolved. A gap between the Kyoto commitment period and subsequent commitment periods is likely because under current rules any new agreement requires ratification by the governments of over 100 countries. Ratification of the Kyoto protocol itself took eight years. The United Nations is considering legal options to support a more rapid ratification.

One consequence of this anticipated gap is uncertainty over the future of the growing trade in carbon emissions. Another is delayed opportunities to address the many shortcomings of the Kyoto Protocol by, for example: improving the mechanisms for setting emissions targets, and for reporting, monitoring, and verifying emissions; setting long-term goals; gaining the participation of the USA, and agreement on commitments by the rapidly industrializing countries; establishing better financial support and mechanisms for supporting mitigation, adaptation, capacity building, technology development and transfer for developing countries.

Negotiations in Copenhagen among the USA, China, India, Brazil, and South Africa created the Copenhagen Accord, a major political, rather than legal, effort to address that potential gap. Although not a binding agreement, the Accord had gained the support of 141 nations by June 2011.

COP16 in Cancún reconfirmed the barriers to rapid progress.

WHAT IS ON OFFER UNDER THE COPENHAGEN ACCORD?

Annex I (industrialized) countries, emissions targets for 2020. Many Annex I countries are proposing doubling or tripling their Kyoto commitments; others are proposing even greater cuts. Some are offering a lower unconditional goal, but a higher goal conditional on the actions of other countries, and the terms of an international agreement. Earlier base years imply higher levels of reduction because populations and economies were smaller.

Country	Base year	Lower commitment	Higher conditional commitment
Canada	2005	–17%	
USA	2005	–17%	
Australia	2000	–5%	–15% to –25%
Kazakhstan	1992	–15%	
Belarus	1990	–5% to –10%	
Croatia	1990	–5%	
European Union	1990	–20%	–30%
Iceland	1990	–30%	
Japan	1990	–25%	
Liechtenstein	1990	–20%	–30%
Monaco	1990	–30%	
New Zealand	1990	–10%	–20%
Norway	1990	–30%	–40%
Russia	1990	–15%	–25%
Switzerland	1990	–20%	–30%
Ukraine	1990	–20%	

Commitments made by largest emitters among developing nations for 2020 emissions targets
Some developing nations (not required to make emission reduction commitments under the Kyoto Protocol) are among the world's largest emitters, and have made substantial voluntary reduction commitments.

	CO_2 emissions		CO_2 emissions per unit of GDP compared to 2005		CO_2 emissions growth trajectory	
Brazil	–36.1% to –39.8%	China	–40% to 45%	South Africa	–34%	
Indonesia	– 26%	India	–20% to 25%	Mexico	–30%	

COPENHAGEN ACCORD
Countries in broad agreement
with the accord
1 April 2011

in broad agreement with accord

expressed intention post-COP15
to be listed as in agreement

COPENHAGEN ACCORD (2009)

- Reduction of emissions with a goal of holding global temperature increases below 2°C, with future consideration of strengthening the goal to 1.5°C.
- Continuation of the Kyoto Protocol.
- Annex I nations to implement quantified economy-wide emissions reductions by 2020.
- Developing nations to implement mitigation actions.
- Enhanced action to support adaptation, and to increase resilience in developing countries.
- Developed countries to address the needs of developing countries with a goal of mobilizing $30bn a year by 2012, and $100bn a year by 2020.

Critics point out that the Accord:
- Is not legally binding.
- Does not set compulsory emissions targets, nor include a long-term vision on emission reductions for 2050.
- Has only weak provisions on mitigation by developing countries.
- Does not define the specifics of monitoring and verification mechanisms.

CANCUN (2010)

- Continues the process of negotiation and confirms much of the Copenhagen Accord, with the addition of a Green Climate Fund.
- Establishes a new Cancún Adaptation Framework.
- Boosts action on deforestation and forest degradation.
- Promotes increased technology cooperation.

BUYERS AND SELLERS

Of CDM and Joint Implementation credits
2009

BUYERS

- other 1%
- Spain, Portugal & Italy 7%
- UK 37%
- Japan 13%
- Germany, Sweden & other Baltic countries 20%
- Netherlands & other Europe 22%

SELLERS

- other 2%
- rest of Latin America 4%
- Brazil 3%
- Africa 7%
- rest of Asia 5%
- India 2%
- Central Asia 5%
- China 72%

The trade in carbon credits is intended to encourage investment in energy efficiency, renewable energy and other ways of reducing emissions, and share the burden of reducing emissions globally.

Under the Kyoto Protocol, countries required to reduce their emissions are entitled to purchase "carbon credits" from developing countries, or from industrialized countries whose emissions are below the level required. The credits cover emissions of all greenhouse gases, expressed as carbon dioxide equivalents (CO_2e).

Carbon markets and exchanges are being established to facilitate the trade. There are three main sources for credits: project-based markets, allowance-based markets, and Assigned Amount Units, which allow Kyoto Annex B countries to sell emissions that were permitted but not "used".

Project-based markets encourage investment in companies or schemes committed to reducing emissions. These are dominated by projects implemented as part of the Clean Development Mechanism (CDM) and Joint Implementation (JI). The main buyers are the industrialized and transition economies of Europe and Japan. Among the developing countries, China was by far the main seller in 2009, followed by Brazil. Africa, once quite marginal to this effort, accounted for 7 percent of 2009 sales. However, carbon markets as a whole still account for less than 3 percent of annual global greenhouse gas emissions.

With the Kyoto Protocol commitments set to expire in 2012, the future of the legal mechanisms supporting many of the market-based initiatives is uncertain. While discussions in Copenhagen in 2009 included plans for improving the Clean Development Mechanism and other modifications to the market-based mechanisms, the lack of international agreement creates uncertainty for the continued development of emissions trading.

PROJECT-BASED MARKETS

Under the Clean Development Mechanism (CDM), buyers invest in projects or companies where there is a commitment to use the money to reduce greenhouse gas emissions. The types of projects funded under the scheme, have changed over time. Much of the initial focus on limiting industrial gas emissions and capturing methane from waste management has abated, and the emphasis is shifting to investments in energy efficiency and renewable energy.

While putting a price on carbon may encourage reductions in emissions, some of these projects have also been criticized by NGOs for their adverse social and environmental impacts as well as insufficient monitoring. Some are viewed as unsustainable, such as creating single-species plantations, large hydroelectric dams, and oil and coal production. Others, such as incineration of landfill materials and landfill gas harvesting, have interfered with more carbon-efficient recycling efforts and the livelihoods of poor people working as wastepickers in many parts of the world. There are also concerns that some projects are playing the system: deliberately producing gases in order to gain the benefits from destroying them.

CDM INVESTMENT

By type of project
2009

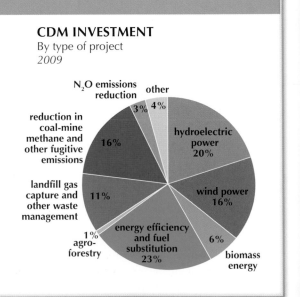

- N_2O emissions reduction 3%
- other 4%
- reduction in coal-mine methane and other fugitive emissions 16%
- hydroelectric power 20%
- landfill gas capture and other waste management 11%
- wind power 16%
- agro-forestry 1%
- energy efficiency and fuel substitution 23%
- biomass energy 6%

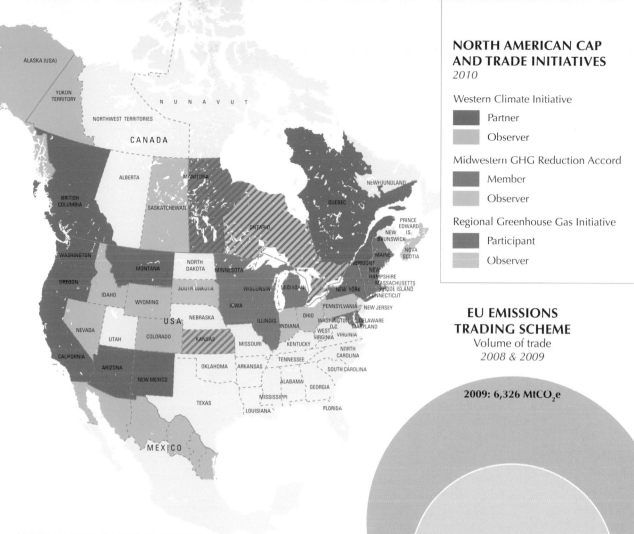

NORTH AMERICAN CAP AND TRADE INITIATIVES
2010

Western Climate Initiative

- Partner
- Observer

Midwestern GHG Reduction Accord

- Member
- Observer

Regional Greenhouse Gas Initiative

- Participant
- Observer

EU EMISSIONS TRADING SCHEME
Volume of trade
2008 & 2009

2009: 6,326 MtCO₂e

2008: 3,093 MtCO₂e

ALLOWANCE-BASED MARKETS

Allowance-based markets enable large companies, such as energy producers, to purchase emissions allowances under schemes administered by regional or international bodies. There are several initiatives in North America: the Regional Greenhouse Gas Initiative, now operational, and two further initiatives under development, although progress has been slow. In addition, New Zealand has created a mandatory economy-wide emissions trading scheme.

The EU Emissions Trading Scheme is, however, by far the largest of these "cap and trade" programs. The volume of trade on the scheme more than doubled between 2008 and 2009 but, due to the recession, the value of the entire market grew by only 6 percent. The decline in price is significant as it reduces the incentive to invest in renewable energy sources and other emission-reduction strategies.

GLOBAL EMISSIONS MARKETS
Change in volume and value
2008 & 2009

- volume (MtCO₂e)
- value (US$)

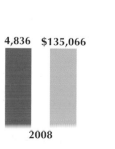

4,836 $135,066

8,700 $143,735

2008   2009

98–99 Financing the Response ▶▶

STATUS OF CLIMATE FUNDS
Value of total funds
2003–April 2011

- pledged
- approved for spending
- disbursed

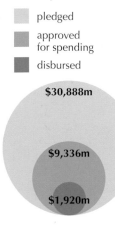

$30,888m

$9,336m

$1,920m

A DROP IN THE OCEAN
Total climate funds as percentage of total official and private Overseas Development Assistance
2003–April 2011

disbursed climate funds
0.07%

pledged climate funds
1.15%

**Total ODA
$2,680,879m**

Finance for responding to climate change has improved dramatically over the past five years, but is still far from what is required.

The Cancún Agreements include $100 billion a year by 2020 for mitigation and adaptation in developing countries. Yet, finance from 24 international climate funds since 2003 is only just over 1 percent of total development assistance.

Developing countries need money to help them meet the challenges of climate change. Funding agreed under the UN Framework Convention on Climate Change is distributed mainly through the World Bank and Global Environment Facility (GEF), which supports action in developing countries on a range of environmental issues. However, bilateral donors (Japan and the UK in particular) and regional banks are gearing up to significant levels.

The private sector is also getting involved. Companies of all sizes are taking stock of their risks and working out how to protect the value of their investments by minimizing the impact of climate change. Several funds are leveraging far greater funds from private financial investments. The expansion of climate funding to include sustainability issues can be seen in the development of green banks and investment funds. The REDD+ (reducing deforestation and degradation) scheme should boost funding for carbon management, although the benefits to local communities remain to be proven.

Tracking the progress of finance is difficult, as many of the funds overlap with environmental and development objectives. The clear commitment to additional funds, beyond the development target of 0.7 percent of GNI, has not been matched by transparent reporting, nor by measurable and effective outcomes. A test case will be the promised $30 billion in fast-start climate finance through 2012.

Countries were initially helped to assess their emissions and institutional capacity. The focus has now shifted to significant actions to reduce emissions, and strategies and measures to adapt to change. The challenges ahead are to increase the scale of effort, particularly in new technologies and technology transfer, while protecting those most vulnerable to adverse climate impacts.

World Bank

The World Bank administers the Climate Investment Fund, providing incentives for integrating climate resilience in 12 pilot countries and regions. It also hosts the GEF, supporting climate change through the GEF Trust Fund, and managing the UNFCCC Least Developed Countries Fund and the Special Climate Change Fund. GEF provides support to the Adaptation Fund Board, and funds some climate-related projects from its own trust funds.

78–79 *Building Capacity to Adapt*; 86–87 *Counting Carbon*; 94–95 *Looking Beyond Kyoto*; 96–97 *Trading Carbon Credits*

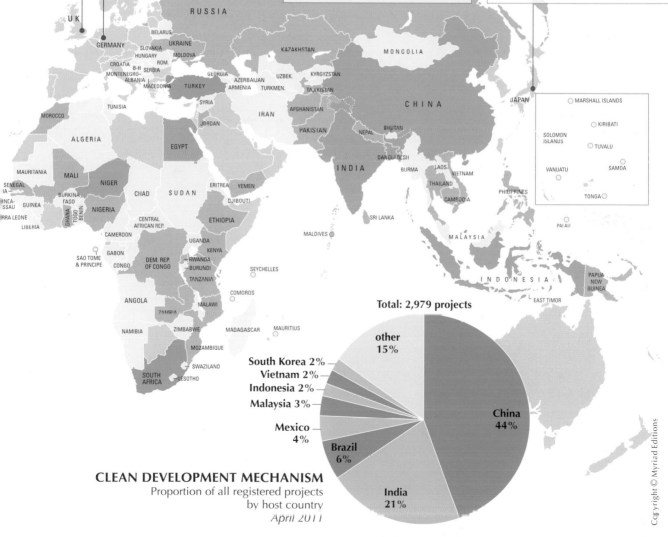

UK finance

The Green Investment Bank, with initial funds of £3bn, will provide high-risk funding for low-carbon infrastructure projects such as off-shore wind farms. The bank will lend from 2012/13 but will only assume powers to borrow on the private market from 2015. The UK has also set a budget of £2.9bn for fast track climate finance for developing countries from 2011–15.

CLIMATE FINANCE
Value of developing-country projects approved (excluding regional and global projects)
2003–April 2011

$200m or more

$100m – $199m

$10m – $99m

less than $10m

not included

Adaptation Fund Board

The Adaptation Fund was established to finance adaptation projects and programmes in developing countries that are Parties to the Kyoto Protocol. Finance is from a surcharge on credits under the Clean Development Mechanism. It is the only climate fund with significant representation by developing countries (the beneficiaries of the funding).

Hatoyama Initiative

The Hatoyama Initiative is a national carbon-regulation scheme, announced at the Copenhagen Summit in December 2009. It targets a 25% cut in global warming emissions below 1990 levels by 2020. Finance is largely public, but over a third of pledges are from the private sector.

Total: 2,979 projects

other 15%

South Korea 2%
Vietnam 2%
Indonesia 2%
Malaysia 3%
Mexico 4%
Brazil 6%
India 21%
China 44%

CLEAN DEVELOPMENT MECHANISM
Proportion of all registered projects by host country
April 2011

Copyright © Myriad Editions

Why not
take a bike?

$2.50 Daily access available at kiosk.

$ annual subscriptions available

rneb share.com.au

13 1,590.

PART 7 Committing to Solutions

Concentrations of greenhouse gases in the atmosphere are continuing to rise. Every day we continue on this current path we are moving further from the reductions in emissions that scientists believe are needed to avoid the most dangerous impacts on ecosystems, agriculture, health, and communities. Action and inaction today will be reflected far into the future.

The remaining uncertainties no longer need to delay concerted effort to reduce emissions. While we cannot forecast with precision the timing, location, and intensity of the consequences of climate change, it is clear that many impacts will be severe and some will result in disasters with tremendous economic losses and incalculable human costs. Arctic communities are retreating from shorelines, and some island communities are migrating. The limits to our ability to cope with current climate extremes have been exposed at tragic cost. We must pursue adaptation to consequences we can no longer avoid, and prepare as best we can for the unanticipated impacts, such as ocean acidification.

Delay comes with substantial risks. The climate system responds slowly, so the benefits of actions will not be realized for decades. The longer we delay, the greater the risks and the more difficult it becomes to stabilize climate change at safe levels. Higher temperatures bring greater risk of crossing over tipping points that shift systems into entirely new, irreversible states. There is risk that an ice-free Arctic might absorb more heat energy and accelerate warming, or that more rapidly melting ice sheets will lead to much higher sea-level rise. As time passes, many adaptations will also be more expensive to implement.

Reducing greenhouse gas emissions does not ensure that the climate and ecosystems will return to the way they were. Some of the worst consequences can be avoided, but only if we achieve much deeper emissions reductions than the international community has currently negotiated.

Motivations for action begin with this awareness, that much is required in the years ahead. Here we all face the tension of our daily lives and our knowledge of the future. We must convey positive images of success and avoid relying solely on the fear of catastrophe. Its correlate is despair and acquiescence in the face of seemingly insurmountable challenges.

Individual efforts are an essential starting point, and knowledge and technologies exist to support substantial improvements, but meeting the scale of this challenge will require more than personal action. The major changes needed to increase the efficiency and resilience of buildings, cities, transportation, energy, and other systems will require long-term vision, leadership, cooperation, innovation, and investment from business and governments at all levels. Assuring their prompt and continued commitment to action is as important as our individual efforts.

> We live in some of the most challenging times that perhaps any generation has faced, but also one of the most exciting moments where the possibilities of re-shaping and re-focusing towards a sustainable 21st century have never been more tangible.

Achim Steiner
UN Under Secretary General and Executive Director, UN Environment Programme (UNEP)

REDUCE, REUSE, RECYCLE

Actions in order of effectiveness:
1. Buy things to last, such as clothes, and appliances, so you buy less.
2. Reuse items, such as shopping bags or water bottles.
3. Recycle what cannot be reused.

These actions cut the amount of GHGs emitted during resource extraction or harvest, manufacturing, transport, and disposal, as well as saving energy and reducing pollution.

People around the world are making lifestyle changes to reduce the greenhouse gas emissions associated with their everyday activities. From the way we run our homes, to the ways we use collective spaces for work, play, and worship, there is plenty of advice on how to live more sustainably.

In addition to benefiting the climate, most of the recommended actions also result in households making long-term financial savings. Some savings are immediate, involving low-cost or no-cost adjustments; others require long-term investment. Many of the decisions taken to save energy will also improve people's quality of life and health.

Together, the actions of millions of people could add up to considerable savings in greenhouse gases, but they will not, on their own, be sufficient to halt climate change. Individuals also need to put pressure on government representatives and companies to take the larger-scale collective action necessary to achieve a reduction in emissions of 60 to 80 percent.

REDUCING EMISSIONS AT HOME

Use energy efficient appliances and lighting.

Insulate and seal ducts and windows.

Insulate wall cavities and loft spaces.

Turn the heating thermostat down, and the air conditioning up.

Use water efficiently. It is very energy intensive to purify and distribute water.

Increase use of renewable energy sources.

Be a carbon-conscious shopper. Buy local, used, and recycled goods.

TRANSPORTATION

Choosing to live near your work and cutting the length of your daily commute will reduce your carbon footprint and save you time, especially if you live close enough to cycle, take public transport, or walk to work.

Purchasing a more energy-efficient car is an effective long-term investment option, but there are many immediate low-cost and no-cost energy saving actions you can take to reduce the emissions of any vehicle.

CAR USE
Possible savings in total US energy use if measures were adopted by all US households
2008

Potential saving: 17.6% of current US total energy use

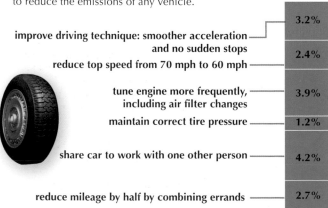

improve driving technique: smoother acceleration and no sudden stops — **3.2%**

reduce top speed from 70 mph to 60 mph — **2.4%**

tune engine more frequently, including air filter changes — **3.9%**

maintain correct tire pressure — **1.2%**

share car to work with one other person — **4.2%**

reduce mileage by half by combining errands — **2.7%**

ENGAGE GOVERNMENT, COMMUNITIES, AND COMPANIES

There are many ways you can influence changes that will have an impact beyond your immediate household or workplace:

- Challenge your political representatives to take climate change seriously.
  Don't ignore the elephant in the room – the message of an artwork created by 3,000 students and teachers at the Ryan International School in New Delhi, and volunteers from the Indian Youth Climate Network. It was photographed by aerial artist Daniel Dancer in November 2010, as part of the 350 EARTH project, which used aerial art to send messages to governments around the world.
- Take part in a day of action in your community, campus, or church.
- Bring your neighbors together to discuss climate change.

- Choose investment funds that support low carbon futures, such as renewable energy sources.
- Ask businesses to disclose any risks they might face due to climate change.
- As a corporate shareholder, encourage sustainability plans that reduce the carbon footprint of production processes and supply chains.
- Encourage your community to incorporate climate change mitigation and adaptation in growth and renewal plans.

ENERGY SAVINGS AT WORK

- Choose energy-efficient equipment and energy-efficient settings.
- Turn off computers and photocopiers at the end of the day.
- Unplug appliances that drain energy when not in use.
- Save paper and always use both sides of the paper.
- Establish an anti-idling policy for the vehicle fleet.
- Use natural lighting when possible.
- Coordinate with vending machine vendor to turn off advertising lights.
- Consider alternative work schedules and telecommuting to reduce GHG from commuting.
- Use coffee cups instead of disposable cups.

CONSIDER BECOMING CARBON NEUTRAL

Calculate your carbon emissions, and make changes that allow you to reduce your emissions as much as possible. Emissions that cannot be reduced can be compensated or offset by paying one of many firms to support renewable energy production or carbon sequestration projects, such as tree planting.

Carbon calculators can be found at the following:
US Environmental Protection Agency: Household Emissions Calculator, Office Carbon Footprint Tool: www.epa.gov
Act on CO2: www.carboncalculator.direct.gov.uk
in running High © www.myhkmpirint.org

Toyota reduced its CO_2 emissions by

20%

per vehicle produced 2001–08

CORPORATE ENERGY REDUCTION SAVES MONEY

Estimates of savings from increased efficiency

total savings

annual savings

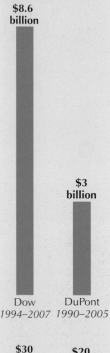

$8.6 billion

$3 billion

Dow
1994–2007

DuPont
1990–2005

$30 million

$20 million

IBM
2007–08

Alcoa
since 2002

The policies, practices, and investments of governments and businesses will have the greatest impact on our future. Individuals can seek to influence those decisions.

Climate change presents a central challenge: how to shift from the current path of social, economic, and technological development to one that reduces emissions and prepares us for future climates. The necessary reduction in emissions of between 60 and 80 percent requires large-scale investment, policy development, and implementation. The control of these areas rests largely with governments and businesses, but the adaptation and mitigation effort requires champions within organizations to lead change. These people, in turn, need the encouragement and support of employees, voters, shareholders, and others to convene the dialogues, identify the risks and opportunities, and promote a broad, long-term vision of sustainability.

Major corporations and government leaders at all levels have already brought about substantial reductions in greenhouse gas emissions, and made timely adaptations to climate change – in some cases in response to citizen and shareholder calls for action. Many of the changes made have provided substantial economic and other benefits. Indeed, the increasing international competition to be the leader in emerging climate-friendly technologies and green jobs, reflects a growing recognition of the multiple advantages of increasing energy efficiency, reducing emissions, and decreasing the risks associated with climate change.

ASSESSING THE RISK TO BUSINESS

Businesses will be affected by climate change both directly and indirectly, not only in terms of their core operations, but the resources they depend on, the parts they buy, the transportation and other services they use, and the social and environmental conditions that make their product attractive to the customer. In today's global markets, companies are likely to feel the impact that climate change is having on the other side of the world.

Aspects of business on which climate change may impact

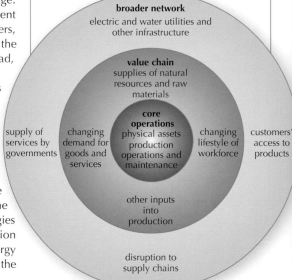

ONLINE TOOLS AND RESOURCES

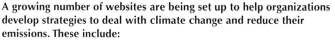

A growing number of websites are being set up to help organizations develop strategies to deal with climate change and reduce their emissions. These include:

- US EPA Energy Star provide customized support for diverse types of enterprises, including government, healthcare, higher education, hospitality/entertainment, retail, small businesses, congregations, service and products providers www.energystar.gov
- UKCIP Business Areas Climate Impacts Assessment Tool (BACLIAT) offers a structured examination of threats and opportunities associated with climate change www.ukcip.org.uk/bacliat
- Global Framework for Climate Risk Disclosure www.unepfi.org
- Cool California supports planning by local governments and CalAdapt leads sectoral adaptation planning: www.CoolCalifornia.org, www.climatechange.ca.gov/

local government

MEETING THE SCALE OF ACTION REQUIRED

Action by governments and larger organizations is necessary to achieving mitigation and adaptation goals. Individual households cannot achieve necessary emissions reductions or adaptations.

- **National Governments** can push for more rapid progress on international emissions reduction commitments and support for adaptation. They can revise laws, regulations, and rules to remove barriers to reducing emissions and increasing adaptation.
- **Local government representatives** are key to implementation of many actions involving construction, land use, infrastructure, and urban design. See Local Governments for Sustainability: www.iclei.org. And, in the absence of national and international action, regional and local action is more important.
- **Workplaces, schools, and places of worship** can adopt numerous strategies to reduce emissions and improve adaptation.
- **Companies and governing bodies** can invest in energy conservation measures, renewable energy, demonstration projects and sharing lessons and innovations. They can also set standards and certifications that allow consumers to make informed choices.
- **Corporations** can develop a greenhouse gas registry, conduct life-cycle analysis to minimize carbon emissions throughout supply chain. They can also assess, report, and improve adaptation to the risks of climate change to operations and financial plans.

PARTICIPATION IN POLICY MAKING

The participation of Non-Governmental Organizations at the UNFCCC Conference of Parties (COP) and Meeting of Parties (MOP) events reflects the growing number of people from all over the world seeking to inform policy on climate change mitigation and adaptation.

The NGOs represent diverse constituencies, including indigenous peoples, business and industry, research and academic institutions, environmental groups, municipal leaders, trade unions, women, and young people. Virtual participation was also available at the recent meetings in Cancún via live and on-demand webcasting, YouTube videos, Facebook, and Twitter.

Number of observer organizations at COP/MOP events
1995–2010

- inter-governmental organization
- non-governmental organization

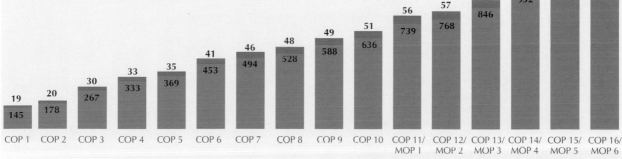

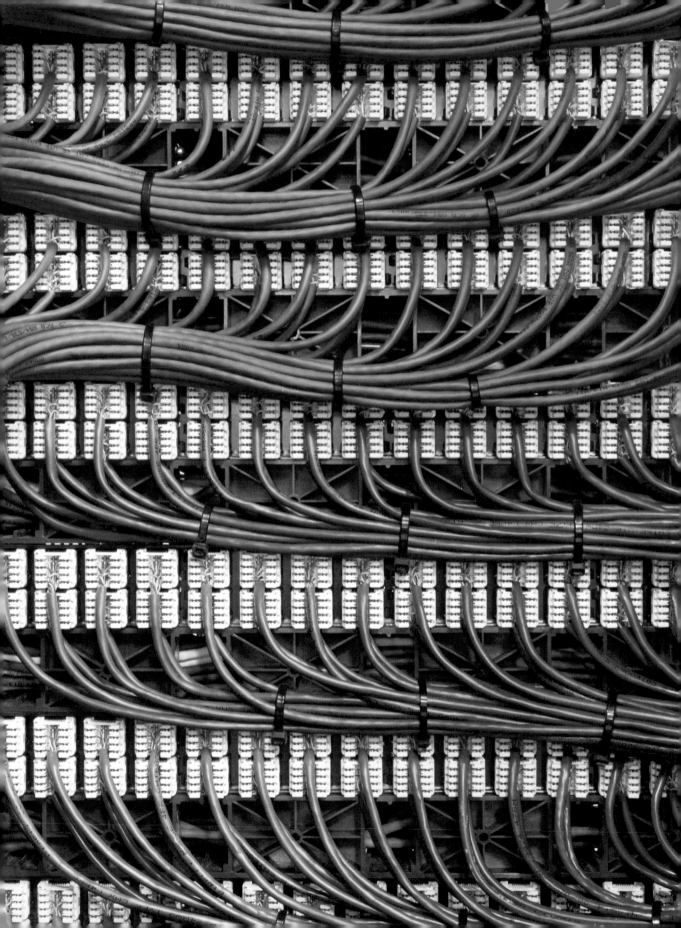

PART 8 Climate Change Data

We respond to what we observe. Data on all aspects of climate change are the bedrock for action: raising awareness leading to understanding, knowledge and wisdom. For an issue of such prominence, it is surprising how difficult it is to gather relevant data, assure its quality and usefulness, and make it widely available and easily accessible. However, there are exciting prospects coming forward too.

Climate change data – across a wide spectrum – come from many different sources. Many of the basic indicators, such as freshwater resources and carbon emissions, are self-reported by countries using agreed methods of accounting. But some countries do not have long monitoring records or do not report regularly. Certain information is more reliable than other information. For instance, it is difficult to monitor the many small household investments in renewable energy, so data are more likely to represent large-scale projects such as hydropower and wind farms. Climate data is sparse in many regions, despite the heavy investment in global climate models. Time series data are often problematic, as measurement standards and technologies have changed.

Most of the global indicators are reported only at the national level. The causes of climate change, potential impacts, and the susceptibility of populations and environmental systems vary greatly within vast countries such as Russia, China, and the USA, and even within smaller countries such as Ghana. For instance, population movements make it difficult to estimate how many people living in a coastal area might be affected by sea-level rise or severe storms, how many living in water-stressed regions are likely to experience greater water stress, or who might be early adopters of household energy-saving appliances.

Access to printed reports and online services remains a challenge for many scientists, decision-makers, and negotiators, especially in developing countries. Power outages interrupt web access; distribution of publications is expensive. And, once found, data must be interrogated to inform decision making. Attempts to build a comprehensive knowledge-base of linked data (using the semantic web) are at an early stage at best.

New horizons for climate-related data promise vast improvements. This third edition of the Atlas has benefited from significant upgrades in the World Resources Institute Climate Analysis Indicators Tool (including free access to some data), coverage of climate-change finance (especially the Climate Funds Update website), and a plethora of portals offering case studies, new reports and advice (from weADAPT.org to cdm-watch.org). As new technology comes online, from smart cell phones to remote sensing, the volume of information is growing exponentially. A good example of what could be more widely available is Google.org's Earth Engine: a digital model of the planet is updated daily to interpret changes in rainforests and water resources.

Making fundamental information more widely and easily available is essential. Actors from local communities to global negotiators would benefit. Facing the challenge of climate change with such information would be far less daunting.

> Ensuring the integrity of research data is essential for advancing scientific, engineering, and medical knowledge, and for maintaining public trust in the research enterprise.

National Academy of Sciences, Committee on Science, Engineering, and Public Policy, 2009

	1 Total population 1,000s 2010	2 GNI per capita PPP US$ 2010	3 Human Development Index 2010	4 Water withdrawn as % of renewable freshwater resources 2008 or latest available	5 Population in coastal cities as % of total in low-elevation coastal zone 2000	6 People at risk from sea-level rise if no adaptation 1,000s projected 2100
Afghanistan	29,117	1,419	0.349	36%	no coastal population	0
Albania	3,169	7,976	0.719	4%	42%	6
Algeria	35,423	8,320	0.677	53%	83%	435
Angola	18,993	4,941	0.403	0%	65%	840
Antigua & Barbuda	89	17,924	–	3%	71%	49
Argentina	40,666	14,603	0.775	4%	92%	76
Armenia	3,090	5,495	0.695	36%	no coastal population	0
Aruba	107	–	–	–	84%	8
Australia	21,512	38,692	0.937	5%	90%	88
Austria	8,387	37,056	0.851	5%	no coastal population	0
Azerbaijan	8,934	8,747	0.713	35%	no coastal population	0
Bahamas	346	25,201	0.784	–	80%	75
Bahrain	807	26,664	0.801	220%	91%	34
Bangladesh	164,425	1,587	0.469	3%	25%	13,282
Barbados	257	21,673	0.788	76%	61%	31
Belarus	9,588	12,926	0.732	7%	no coastal population	0
Belgium	10,698	34,873	0.867	34%	96%	12
Belize	313	5,693	0.694	1%	65%	13
Benin	9,212	1,499	0.435	0%	75%	0
Bhutan	708	5,607	–	0%	no coastal population	0
Bolivia	10,031	4,357	0.643	0%	no coastal population	0
Bosnia & Herzegovina	3,760	8,222	0.71	1%	0%	0
Botswana	1,978	13,204	0.633	2%	no coastal population	0
Brazil	195,423	10,607	0.699	1%	89%	2,795
Brunei	407	49,915	0.805	1%	80%	3
Bulgaria	7,497	11,139	0.743	29%	89%	4
Burkina Faso	16,287	1,215	0.305	8%	no coastal population	0
Burma	50,496	1,596	0.451	3%	37%	1,459
Burundi	8,519	402	0.282	2%	no coastal population	0
Cambodia	15,053	1,868	0.494	0%	9%	10
Cameroon	19,958	2,197	0.46	0%	85%	1,938
Canada	33,890	38,668	0.888	2%	85%	467
Cape Verde	513	3,306	0.534	7%	43%	57
Central African Republic	4,506	758	0.315	0%	no coastal population	0
Chad	11,506	1,067	0.295	1%	no coastal population	0
Chile	17,135	13,561	0.783	1%	77%	400
China	1,354,146	7,258	0.663	20%	54%	15,059
Colombia	46,300	8,589	0.689	1%	78%	335
Comoros	691	1,176	0.428	1%	39%	31
Congo	3,759	3,258	0.489	0%	89%	138
Congo, Dem Rep.	67,827	291	0.239	0%	37%	1
Cook Islands	20	–	–	–	48%	–
Costa Rica	4,640	10,870	0.725	2%	41%	8
Côte d'Ivoire	21,571	1,625	0.397	2%	88%	0
Croatia	4,410	16,389	0.767	1%	39%	10
Cuba	11,204	–	–	20%	71%	250
Cyprus	880	21,962	0.81	19%	86%	14
Czech Republic	10,411	22,678	0.841	15%	no coastal population	0
Denmark	5,481	36,404	0.866	11%	74%	22
Djibouti	879	2,471	0.402	6%	95%	86

7 CO₂ emissions			8 Methane emissions			9 Carbon intensity		
Per capita	Total from burning of fossil fuels	Total from transportation	Per capita CO₂e	Total	Total from agriculture (plus N₂O)	Tonnes of CO₂ emitted per $1,000 GDP		
tonnes 2008	million tonnes 2008		2005	million tonnes CO₂e 2008		2008	1998– 2008	
–	–	–	0.4	10.7	–	–	–	Afghanistan
1.23	4	2	0.9	2.9	0.68	0.15	3	Albania
2.56	88	18	0.8	27.6	1.17	0.05	16	Algeria
0.59	11	3	2.7	44.7	0.43	–0.04	–	Angola
–	–	–	0.5	0.0	–	–	–	Antigua& Barbuda
4.36	174	40	2.5	95.1	0.44	–0.02	139	Argentina
1.71	5	1	0.9	2.7	1.12	–0.77	1	Armenia
–	–	–	–	–	–	–	–	Aruba
18.48	398	68	6.3	128.9	0.77	–0.09	110	Australia
8.31	69	21	1.0	8.1	0.31	0.05	8	Austria
3.37	29	5	1.7	14.0	1.58	–4.69	5	Azerbaijan
–	–	–	0.6	0.2	–	–	–	Bahamas
29.08	22	3	3.8	2.8	1.71	–0.19	–	Bahrain
0.29	46	5	0.3	53.6	0.63	0.07	80	Bangladesh
–	–	–	0.4	0.1	–	–	–	Barbados
6.63	64	4	1.4	13.7	2.67	–2.42	13	Belarus
10.36	111	27	0.9	9.1	0.41	–0.15	12	Belgium
–	–	–	0.7	0.2	–	–	–	Belize
0.38	3	2	0.5	3.9	1.05	0.46	–	Benin
–	–	–	1.6	1.0	–	–	–	Bhutan
1.33	13	4	3.6	33.1	1.13	0.10	46	Bolivia
5.18	20	3	0.7	2.6	2.32	0.09	–	Bosnia & Herzegovina
2.37	5	2	2.4	4.5	0.53	–0.20	–	Botswana
1.90	365	135	2.1	389.1	0.43	–0.03	591	Brazil
18.87	7	1	15.6	5.8	1.09	0.29	–	Brunei
6.40	49	8	1.3	10.3	2.49	–1.67	7	Bulgaria
–	–	–	0.8	10.7	–	–	–	Burkina Faso
0.24	12	3	1.5	74.9	0.61	–0.49	78	Burma
–	–	–	0.2	1.8	–	–	–	Burundi
0.31	5	1	1.0	13.5	0.61	–0.10	16	Cambodia
0.23	4	2	1.0	18.5	0.32	0.03	–	Cameroon
16.53	551	127	3.2	102.0	0.63	–0.13	73	Canada
–	–	–	0.3	0.1	–	–	–	Cape Verde
–	–	–	5.9	24.2	–	–	–	Central African Republic
–	–	–	1.2	12.0	–	–	–	Chad
4.35	73	17	0.9	14.8	0.70	–0.06	15	Chile
4.92	6,550	334	0.7	853.3	2.30	–0.41	1,113	China
1.35	60	22	1.4	62.3	0.45	–0.22	89	Colombia
–	–	–	0.3	0.2	–	–	–	Comoros
0.41	1	1	1.6	5.5	0.34	0.21	–	Congo
0.04	3	1	1.0	58.0	0.45	0.01	75	Congo, Dem Rep.
–	–	–	0.2	0.0	–	–	–	Cook Islands
1.45	7	4	0.6	2.6	0.28	–0.03	–	Costa Rica
0.31	6	1	0.5	10.4	0.59	0.15	–	Côte d'Ivoire
4.72	21	6	0.9	4.1	0.69	–0.18	4	Croatia
2.71	31	1	0.8	9.5	0.70	–0.29	–	Cuba
9.49	8	2	0.7	0.6	0.62	–0.07	–	Cyprus
11.20	117	17	1.0	10.5	1.48	–0.70	8	Czech Republic
8.82	48	13	1.0	5.6	0.27	–0.11	10	Denmark
			0.6	0.5				Djibouti

	1 Total population 1,000s 2010	2 GNI per capita PPP US$ 2010	3 Human Development Index 2010	4 Water withdrawn as % of renewable freshwater resources 2008 or latest available	5 Population in coastal cities as % of total in low-elevation coastal zone 2000	6 People at risk from sea-level rise if no adaptation 1,000s projected 2100
Dominica	67	8,549	–	–	35%	47
Dominican Republic	10,225	8,273	0.663	17%	49%	22
East Timor	1,171	5,303	0.502	–	5%	22
Ecuador	13,775	7,931	0.695	4%	71%	398
Egypt	84,474	5,889	0.62	119%	44%	1,970
El Salvador	6,194	6,498	0.659	5%	10%	35
Equatorial Guinea	693	22,218	0.538	0%	50%	0
Eritrea	5,224	643	–	9%	27%	7
Estonia	1,339	17,168	0.812	14%	78%	3
Ethiopia	84,976	992	0.328	5%	no coastal population	0
Fiji	854	4,315	0.669	0%	64%	43
Finland	5,346	33,872	0.871	1%	69%	1
France	62,637	34,341	0.872	15%	77%	128
French Polynesia	272	–	–	–	21%	61
Gabon	1,501	12,747	0.648	0%	37%	24
Gambia	1,751	1,358	0.39	1%	67%	92
Georgia	4,219	4,902	0.698	3%	70%	42
Germany	82,057	35,308	0.885	21%	80%	129
Ghana	24,333	1,385	0.467	2%	65%	0
Greece	11,183	27,580	0.855	13%	73%	240
Grenada	104	7,998	–	–	40%	19
Guatemala	14,377	4,694	0.56	3%	15%	20
Guinea	10,324	953	0.34	1%	74%	10
Guinea-Bissau	1,647	538	0.289	1%	36%	170
Guyana	761	3,302	0.611	1%	60%	120
Haiti	10,188	949	0.404	9%	57%	85
Honduras	7,616	3,750	0.604	1%	51%	40
Hungary	9,973	17,472	0.805	5%	no coastal population	0
Iceland	329	22,917	0.869	0%	53%	6
India	1,214,464	3,337	0.519	40%	50%	16,974
Indonesia	232,517	3,957	0.6	6%	55%	7,596
Iran	75,078	11,764	0.702	68%	49%	131
Iraq	31,467		–	87%	62%	350
Ireland	4,589	33,078	0.895	2%	73%	82
Israel	7,285	27,831	0.872	102%	93%	7
Italy	60,098	29,619	0.854	24%	75%	66
Jamaica	2,730	7,207	0.688	6%	64%	11
Japan	126,995	34,692	0.884	21%	95%	150
Jordan	6,472	5,956	0.681	99%	58%	0
Kazakhstan	15,753	10,234	0.714	29%	no coastal population	0
Kenya	40,863	1,628	0.47	9%	17%	280
Kiribati	100	3,715	–	–	no data	71
Korea, North	23,991	–	–	455%	39%	545
Korea, South	48,501	29,518	0.877	9%	71%	209
Kuwait	3,051	55,719	0.771	2465%	35%	97
Kyrgyzstan	5,550	2,291	0.598	44%	no coastal population	0
Laos	6,436	2,321	0.497	1%	no coastal population	0
Latvia	2,240	12,944	0.769	1%	85%	9
Lebanon	4,255	13,475	–	28%	87%	186
Lesotho	2,084	2,021	0.427	2%	no coastal population	0

Per capita	Total from burning of fossil fuels	Total from transportation	Per capita CO_2e	Total	Total from agriculture (plus N_2O)	Tonnes of CO_2 emitted per $1,000 GDP		
tonnes 2008	million tonnes 2008		2005	million tonnes CO_2e 2008		2008	1998–2008	
–	–	–	0.5	0.0	–	–	–	Dominica
1.99	20	4	0.6	5.9	0.54	–0.20	–	Dominican Republic
–	–	–	–	–	–	–	–	East Timor
1.92	26	12	1.2	15.5	1.10	–0.15	12	Ecuador
2.13	174	35	0.5	38.0	1.20	0.07	27	Egypt
0.95	6	2	0.5	3.1	0.35	–0.07	–	El Salvador
–	–	–	9.9	6.0	–	–	–	Equatorial Guinea
0.09	0	0	0.5	2.4	0.61	–0.18	–	Eritrea
13.14	18	2	1.9	2.5	1.86	–1.23	1	Estonia
0.08	7	4	0.7	54.3	0.45	0.04	55	Ethiopia
–	–	–	0.8	0.7	–	–	–	Fiji
10.65	57	12	1.0	5.2	0.37	–0.14	5	Finland
5.74	368	119	1.0	60.9	0.24	–0.07	103	France
–	–	–	–	–	–	–	–	French Polynesia
2.06	3	1	6.0	8.2	0.50	0.21	–	Gabon
–	–	–	0.4	0.6	–	–	–	Gambia
1.08	5	2	0.8	3.5	0.86	–0.71	2	Georgia
9.79	804	140	0.8	68.5	0.38	–0.09	84	Germany
0.31	7	3	0.4	8.5	0.96	–0.24	–	Ghana
8.31	93	19	0.9	9.7	0.54	–0.17	12	Greece
–	–	–	0.3	0.0	–	–	–	Grenada
0.78	11	5	0.7	8.3	0.41	–0.03	–	Guatemala
–	–	–	1.0	9.4	–	–	–	Guinea
–	–	–	0.7	1.0	–	–	–	Guinea-Bissau
–	–	–	1.7	1.3	–	–	–	Guyana
0.24	2	1	0.4	3.9	0.58	0.23	–	Haiti
1.08	8	3	0.8	5.2	0.74	0.12	–	Honduras
5.28	53	13	1.1	10.9	0.86	–0.46	12	Hungary
6.89	2	1	1.7	0.5	0.18	–0.08	1	Iceland
1.25	1,428	121	0.5	547.7	1.73	–0.41	403	India
1.69	385	69	0.8	183.0	1.56	0.07	132	Indonesia
7.02	505	110	1.4	95.7	3.15	0.16	37	Iran
3.45	97	30	0.5	13.1	4.25	0.88	10	Iraq
9.85	44	13	2.9	11.9	0.32	–0.15	19	Ireland
8.63	63	10	1.7	11.6	0.39	–0.05	3	Israel
7.18	430	110	0.6	34.6	0.37	–0.04	41	Italy
4.44	12	2	0.5	1.2	1.17	0.11	–	Jamaica
9.02	1,151	203	0.2	20.9	0.22	–0.03	35	Japan
3.12	18	5	0.4	2.2	1.31	–0.42	1	Jordan
12.86	202	13	1.8	26.9	5.41	–2.37	17	Kazakhstan
0.22	9	3	0.6	20.1	0.48	0.00	–	Kenya
–	–	–	0.1	0.0	–	–	–	Kiribati
2.91	69	1	1.4	33.7	5.97	0.59	9	Korea, North
10.31	501	79	0.7	33.4	0.67	–0.11	18	Korea, South
25.47	69	11	3.8	9.8	1.04	–0.25	0	Kuwait
1.12	6	1	0.7	3.7	2.99	–1.53	2	Kyrgyzstan
–	–	–	2.2	12.8	–	–	14	Laos
3.49	8	3	0.8	1.9	0.58	–0.56	2	Latvia
3.68	15	4	0.2	1.0	0.64	–0.27	–	Lebanon
–	–	–	0.3	0.5	–	–	–	Lesotho

	1 Total population 1,000s 2010	2 GNI per capita PPP US$ 2010	3 Human Development Index 2010	4 Water withdrawn as % of renewable freshwater resources 2008 or latest available	5 Population in coastal cities as % of total in low-elevation coastal zone 2000	6 People at risk from sea-level rise if no adaptation 1,000s projected 2100
Liberia	4,102	320	0.3	0%	63%	0
Libya	6,546	17,068	0.755	718%	94%	39
Lithuania	3,255	14,824	0.783	10%	82%	3
Luxembourg	492	51,109	0.852	13%	no coastal population	0
Macedonia	2,043	9,487	0.701	2%	no coastal population	0
Madagascar	20,146	953	0.435	4%	38%	141
Malawi	15,692	911	0.385	6%	no coastal population	0
Malaysia	27,914	13,927	0.744	2%	71%	456
Maldives	314	5,408	0.602	16%	2%	179
Mali	13,323	1,171	0.309	7%	no coastal population	0
Malta	410	21,004	0.815	71%	76%	21
Marshall Islands	63	–	–	–	32%	44
Mauritania	3,366	2,118	0.433	14%	27%	20
Mauritius	1,297	13,344	0.701	26%	84%	0
Mexico	110,645	13,971	0.75	37%	61%	675
Micronesia (Fed. States of)	111	3,266	0.614	–	9%	13
Moldova	3,576	3,149	0.623	17%	54%	0
Mongolia	2,701	3,619	0.622	1%	no coastal population	0
Montenegro	626	12,491	0.769	–	65%	–
Morocco	32,381	4,628	0.567	43%	86%	1,820
Mozambique	23,406	854	0.284	0%	46%	4,348
Namibia	2,212	6,323	0.606	2%	85%	1
Nauru	10	–	–	–	72%	4
Nepal	29,853	1,201	0.428	5%	no coastal population	0
Netherlands	16,653	40,658	0.89	12%	86%	97
Netherlands Antilles	201	–	–	–	28%	65
New Zealand	4,303	25,438	0.907	1%	83%	31
Nicaragua	5,822	2,567	0.565	1%	27%	8
Niger	15,891	675	0.261	7%	no coastal population	0
Nigeria	158,259	2,156	0.423	4%	58%	210
Niue	1	–	–	–	0%	–
Norway	4,855	58,810	0.938	50%	59%	10
Oman	2,905	25,653	–	87%	93%	23
Pakistan	184,753	2,678	0.49	1%	54%	829
Palau	21	–	–	–	73%	4
Palestinian Territories	4,409	4,570	–	81%	89%	10
Panama	3,508	13,347	0.755	0%	65%	82
Papua New Guinea	6,888	2,227	0.431	0%	36%	102
Paraguay	6,460	4,585	0.64	0%	no coastal population	0
Peru	29,496	8,424	0.723	1%	47%	357
Philippines	93,617	4,002	0.638	17%	51%	1,796
Poland	38,038	17,803	0.795	19%	73%	107
Portugal	10,732	22,105	0.795	12%	76%	51
Puerto Rico	3,998	–	–	14%	88%	60
Qatar	1,508	79,426	0.803	16%	88%	9
Réunion	837	–	–	–	46%	4
Romania	21,190	12,844	0.767	3%	71%	66
Russia	140,367	15,258	0.719	1%	69%	412
Rwanda	10,277	1,190	0.385	2%	no coastal population	0
Saint Kitts & Nevis	52	14,196	–	–	70%	33

Per capita	Total from burning of fossil fuels	Total from transportation	Per capita CO$_2$e	Total	Total from agriculture (plus N$_2$O)	\multicolumn{2}{c}{Tonnes of CO$_2$ emitted per \$1,000 GDP}	
7 CO$_2$ emissions			**8** Methane emissions			**9** Carbon intensity		
tonnes *2008*	million tonnes *2008*		*2005*	million tonnes CO$_2$e *2008*		*2008*	*1998–2008*	
–	–	–	0.3	0.9	–	–	–	Liberia
7.15	45	10	2.5	14.6	0.85	–0.29	–	Libya
4.24	14	5	1.2	3.9	0.71	–0.65	2	Lithuania
21.27	10	6	1.2	0.6	0.38	–0.03	1	Luxembourg
4.40	9	1	1.0	2.0	2.02	–0.91	1	Macedonia
–	–	–	1.0	17.3	–	–	–	Madagascar
–	–	–	0.2	3.2	–	–	–	Malawi
6.70	181	41	1.8	46.1	1.30	0.11	–	Malaysia
–	–	–	0.1	0.0	–	–	–	Maldives
–	–	–	1.1	13.4	–	–	–	Mali
6.23	3	1	0.4	0.2	0.58	–0.09	–	Malta
–	–	–	–	–	–	–	–	Marshall Islands
–	–	–	1.7	5.1	–	–	–	Mauritania
–	–	–	0.2	0.3	–	–	–	Mauritius
3.83	408	147	1.8	184.8	0.53	–0.05	77	Mexico
–	–	–	–	–	–	–	–	Micronesia (Fed. States of)
1.95	7	1	0.7	2.7	3.37	–3.92	2	Moldova
4.33	11	1	3.0	7.5	5.87	–2.09	20	Mongolia
–	–	–	1.0	7.4	–	–	–	Montenegro
1.35	42	11	0.3	10.5	0.76	0.03	–	Morocco
0.09	2	1	0.6	12.6	0.24	–0.05	–	Mozambique
1.86	4	2	2.5	5.1	0.69	0.13	–	Namibia
–	–	–	0.2	0.0	–	–	–	Nauru
0.12	3	1	0.9	25.6	0.46	0.02	32	Nepal
10.82	178	34	1.1	17.3	0.40	–0.09	19	Netherlands
22.91	4	1		–	3.33	0.10	–	Netherlands Antilles
7.74	33	13	6.6	27.3	0.50	–0.05	38	New Zealand
0.73	4	1	1.1	6.0	0.81	–0.13	–	Nicaragua
–	–	–	0.3	4.5	–	–	–	Niger
0.35	52	25	1.1	150.5	0.71	0.05	115	Nigeria
–	–	–	0.4	0.0	–	–	–	Niue
7.89	38	10	1.2	5.4	0.19	–0.04	5	Norway
12.54	35	5	6.8	17.8	1.13	0.34	–	Oman
0.81	134	31	0.6	97.7	1.19	0.12	79	Pakistan
–	–	–	0.2	0.0	–	–	–	Palau
–	–	–	–	–	–	–	–	Palestinian Territories
1.92	7	1	1.0	3.2	0.34	–0.16	–	Panama
–	–	–	0.3	1.6	–	–	–	Papua New Guinea
0.59	4	3	2.6	15.3	0.39	–0.15	–	Paraguay
1.21	35	13	0.7	20.6	0.41	–0.07	36	Peru
0.80	72	20	0.5	41.2	0.65	–0.38	39	Philippines
7.84	299	43	1.2	46.4	1.26	–0.73	27	Poland
4.94	52	18	0.8	8.6	0.43	–0.08	9	Portugal
–	–	–	–	–	–	–	–	Puerto Rico
42.09	54	9	17.7	15.7	1.43	–0.04	–	Qatar
–	–	–	–	–	–	–	–	Réunion
4.18	90	14	1.2	26.6	1.47	–1.16	14	Romania
11.24	1,594	132	2.2	314.5	3.71	–2.75	118	Russia
–	–	–	0.2	2.1	–	–	–	Rwanda
–	–	–	0.7	0.0	–	–	–	Saint Kitts and Nevis

	1 Total population 1,000s 2010	2 GNI per capita PPP US$ 2010	3 Human Development Index 2010	4 Water withdrawn as % of renewable freshwater resources 2008 or latest available	5 Population in coastal cities as % of total in low-elevation coastal zone 2000	6 People at risk from sea-level rise if no adaptation 1,000s projected 2100
Saint Lucia	174	8,652	–	–	62%	10
Saint Vincent & the Grenadines	109	8,535	–	–	0%	20
Samoa	179	4,126	–	–	–	12
São Tomé & Príncipe	165	1,918	0.488	0%	77%	0
Saudi Arabia	26,246	24,726	0.752	943%	96%	60
Senegal	12,861	1,816	0.411	6%	73%	489
Serbia	9,856	10,449	0.735	–	65%	–
Seychelles	85	19,128	–	–	51%	20
Sierra Leone	5,836	809	0.317	0%	37%	45
Singapore	4,837	48,893	0.846	32%	91%	5
Slovakia	5,412	21,658	0.818	1%	no coastal population	0
Slovenia	2,025	25,857	0.828	3%	86%	0
Solomon Islands	536	2,172	0.494	–	42%	45
Somalia	9,359	437	–	22%	28%	98
South Africa	50,492	9,812	0.597	25%	88%	69
Spain	45,317	29,661	0.863	29%	89%	77
Sri Lanka	20,410	4,886	0.658	25%	43%	251
Sudan	43,192	2,051	0.379	58%	20%	0
Suriname	524	7,093	0.646	1%	91%	69
Swaziland	1,202	5,132	0.498	23%	no coastal population	0
Sweden	9,293	36,936	0.885	2%	81%	3
Switzerland	7,595	39,849	0.874	5%	no coastal population	0
Syrian Arab Republic	22,505	4,760	0.589	100%	79%	15
Taiwan	22,900	–	–	–	80%	42
Tajikistan	7,075	2,020	0.58	9%	no coastal population	0
Tanzania	45,040	1,344	0.398	75%	49%	2,083
Thailand	68,139	8,001	0.654	16%	76%	967
Togo	6,780	844	0.428	1%	94%	0
Tonga	104	4,038	0.677	–	58%	64
Trinidad & Tobago	1,344	24,233	0.736	6%	76%	59
Tunisia	10,374	7,979	0.683	62%	84%	264
Turkey	75,705	13,359	0.679	19%	77%	1,089
Turkmenistan	5,177	7,052	0.669	101%	no coastal population	0
Tuvalu	10	–	–	–	0%	6
Uganda	33,796	1,224	0.422	0%	no coastal population	0
Ukraine	45,433	6,535	0.71	28%	65%	404
United Arab Emirates	4,707	58,006	0.815	2032%	95%	20
United Kingdom	61,899	35,087	0.849	5%	83%	253
United States of America	317,641	47,094	0.902	16%	89%	1,589
Uruguay	3,372	13,808	0.765	3%	89%	8
Uzbekistan	27,794	3,085	0.617	118%	no coastal population	0
Vanuatu	246	3,908	–	–	0%	14
Venezuela	29,044	11,846	0.696	1%	86%	1,016
Vietnam	89,029	2,995	0.572	9%	30%	13,815
Yemen	24,256	2,387	0.439	169%	73%	68
Zambia	13,257	1,359	0.395	2%	no coastal population	0
Zimbabwe	12,644	176	0.14	21%	no coastal population	0

Per capita	Total from burning of fossil fuels	Total from transportation	Per capita CO_2e	Total	Total from agriculture (plus N_2O)	Tonnes of CO_2 emitted per $1,000 GDP		
tonnes 2008	million tonnes 2008		2005	million tonnes CO_2e 2008		2008	1998–2008	
–	–	–	0.2	0.0	–	–	–	Saint Lucia
–	–	–	0.3	0.0	–	–	–	Saint Vincent & the Grenadines
–	–	–	0.7	0.1	–	–	–	Samoa
–	–	–	0.2	0.0	–	–	–	São Tomé & Príncipe
15.79	389	94	1.2	27.7	1.54	0.25	12	Saudi Arabia
0.42	5	2	0.8	9.1	0.78	0.04	12	Senegal
6.70	49	6	1.0	7.4	3.55	−1.25	–	Serbia
–	–	–	0.3	0.0	–	–	–	Seychelles
–	–	–	0.4	1.8	–	–	–	Sierra Leone
9.16	44	7	0.4	1.7	0.33	0.19	0	Singapore
6.70	36	6	0.9	4.7	1.10	−0.88	6	Slovakia
8.27	17	6	1.0	2.1	0.60	−0.22	2	Slovenia
–	–	–	3.0	1.4	–	–	–	Solomon Islands
–	–	–	–	–	–	–	–	Somalia
6.93	337	42	1.2	55.3	1.84	−0.64	42	South Africa
6.97	318	95	0.8	36.6	0.43	−0.04	46	Spain
0.61	12	5	0.5	10.2	0.51	−0.06	–	Sri Lanka
0.29	12	7	1.7	65.3	0.55	0.11	–	Sudan
–	–	–	1.3	0.7	–	–	–	Suriname
–	–	–	1.0	1.2	–	–	–	Swaziland
4.96	46	22	0.6	5.7	0.15	−0.10	9	Sweden
5.67	44	17	0.5	3.6	0.15	−0.03	5	Switzerland
2.56	54	12	0.7	12.5	1.99	−0.32	–	Syrian Arab Republic
11.53	264	33	0.4	8.1	0.63	−0.03	–	Taiwan
0.44	3	0	0.5	3.0	1.81	−1.78	2	Tajikistan
0.14	6	3	0.8	30.2	0.38	0.09	–	Tanzania
3.41	229	51	1.4	91.6	1.29	−0.04	89	Thailand
0.17	1	1	0.4	2.7	0.70	0.07	–	Togo
–	–	–	0.6	0.1	–	–	–	Tonga
28.37	38	2	7.5	9.9	2.58	0.34	–	Trinidad & Tobago
2.01	21	5	0.8	8.0	0.73	−0.21	–	Tunisia
3.71	264	40	1.5	105.3	0.70	0.01	76	Turkey
9.41	47	3	10.1	49.1	5.52	−9.07	2	Turkmenistan
–	–	–	–	–	–	–	–	Tuvalu
–	–	–	0.5	13.4	–	–	20	Uganda
6.69	310	24	3.3	153.4	5.79	−4.49	45	Ukraine
32.77	147	25	10.1	41.3	1.19	−0.04	1	United Arab Emirates
8.32	511	115	0.8	46.2	0.29	−0.09	48	United Kingdom
18.38	5,596	1,456	1.8	521.0	0.48	−0.13	442	United States of America
2.29	8	3	6.0	19.9	0.26	0.02	35	Uruguay
4.21	115	5	2.1	53.8	5.01	−4.22	21	Uzbekistan
–	–	–	1.2	0.3	–	–	–	Vanuatu
5.21	146	45	3.2	83.8	0.87	−0.25	52	Venezuela
1.19	103	23	0.8	68.6	1.85	0.40	65	Vietnam
0.95	22	6	0.3	6.6	1.70	0.44	–	Yemen
0.13	2	0	1.6	18.6	0.33	−0.35	–	Zambia
0.70	9	1	0.8	9.5	1.86	0.25	–	Zimbabwe

Sources & Notes

Definition of Key Terms
The glossary published by the Intergovernmental Panel on Climate Change is the principal source for the technical definitions: www.ipcc.ch/pub/gloss.pdf and www.ipcc.ch/pub/syrgloss.pdf.
Other sources for technical terms include:
American Meteorological Society www.ametsoc.org
California Climate Change Portal www.climatechange.ca.gov/glossary
Glossary of Meteorology amsglossary. allenpress.com
National Geographysical Data Center: www.ngdc.noaa.gov

Forewords
Wangari Maathai
www.Greenbeltmovement.org
Philippe Cousteau
WWF. Living Planet Report 2010: Biodiversity, biocapacity and development. Gland: WWF International; 2010
www.footprintnetwork.org

Part 1: Signs of Change
World Meteorological Organization. WMO statement on the status of the global climate in 2010. Geneva: World Meteorological Organization; 2011. 2011. WMO-No.1074.
"Climate change has a taste…"
Atiq Rahman, in closing keynote at climate adaptation conference, Dhaka. 2011, March 31. One World Group http://oneworldgroup.org

1 Warning Signs
International Panel on Climate Change (IPCC) Fourth Assessment Report, Working Group I. The physical science basis. Summary for policymakers. IPCC; 2007. www.ipcc.ch
International Panel on Climate Change (IPCC) Fourth Assessment Report, Working Group II. Climate change 2007: Impacts, adaptation and vulnerability. Summary for policymakers, IPCC; 2007. www.ipcc.ch
International Panel on Climate Change (IPCC) Fourth Assessment Report, Working Group III. Mitigation of climate change. Summary for policymakers. IPCC; 2007. www.ipcc.ch
All three global temperature records agree that 2010 was tied as the warmest year on record, since 1850. The tie is with either 1998 or 2005, depending on which time series is used. Differences between 1998 and 2005 are only a few hundredths of a degree.
In April 2007, the IPCC stated with "high confidence" that recent warming has affected terrestrial, marine and freshwater biological systems, glaciers and rivers. Based on an analysis of over 29,000 data sets, contained in 75 studies from around the world, it concluded that over 90 percent of observed changes were consistent with climate change.
RECORD HIGHS
Countries that set all-time heat records in 2010. Graphic by Climate Central. Data from Jeff Masters/Weather Underground. Countries that set all-time heat records in 2010 www.climatecentral.org

OBSERVED CLIMATE CHANGE IMPACTS
IPCC. Fourth Assessment Report, Working Group II. 2007. Figure 1.9. www.ipcc.ch
Note: The cases of significant changes in observations of physical systems (snow, ice and frozen ground; hydrology; coastal processes) and biological systems (terrestrial, marine and freshwater biological systems) met the following criteria: (1) ending in 1990 or later; (2) spanning a period of at least 20 years; (3) showing a significant change in either direction. The paucity of observed data for Africa and Australia is apparent.
Parmesan C, Yohe G. A globally coherent fingerprint of climate change impacts across natural systems. Nature 2003;421:37-42.
Walther G-R. Community and ecosystem responses to recent climate change. Phil Trans R Soc B 2010;365(1549):2019-2024.
USA: heat wave
NOAA: US experienced above average temperatures, rainfall in September. 2010 Oct 7. www.noaanews.noaa.gov
US Climate Extremes Index, NOAA www.ncdc.noaa.gov/extremes/cei/
Russia: state of emergency
NOAA. State of the Climate Report. August 2010 www.ncdc.noaa.gov
China: floods
Rainstorms hit China's northeast as south bathes in heat. People's Daily Online. 2010 Aug 14. http://english.peopledaily.com.cn
Floods, landslides leave 3,185 dead in China this year. 2010 Aug 31. http://news.xinhuanet.com
Death toll in flood-stricken south China nears 400. 2010 June 26. www.reuters.com
Pakistan: floods
Slingo J. Pakistan floods and extreme weather in August 2010. UK Met Office. www.metoffice.gov.uk
National Disaster Management Authority (Pakistan). Summary of damages 2010. www.pakistanfloods.pk
Preliminary damage estimates for Pakistani flood events. Ball State University - Center for Business and Economic Research. 2010 August.
NASA Earth Observatory. MODIS Rapid Response Team.
Australia: floods
Special Climate Statement 24. Australian Government Bureau of Meteorology. 2011 January 25.
Queensland Police media release 2011 Jan 24. www.police.qld.gov.au
Counting cost of Queensland floods. 2011 Jan 15. www.news.com.au

2 Polar Changes
Lemke P et al. Observations: changes in snow, ice and frozen ground. In: Solomon S et al. editors. Climate change 2007: the physical science basis. Contribution of Working Group I to the Fourth Assessment Report of the Intergovernmental Panel on Climate Change. Cambridge, UK and New York, NY, USA: Cambridge University Press; 2007. pp 337-83.
Perovich D et al. Sea Ice Cover 2010 www.arctic.noaa.gov [Accessed 2011 May 22]

Kwok R, Rothrock DA. Decline in Arctic sea ice thickness from submarine and ICESat records: 1958-2008. J Geophys Res Lett 2009;36. L15501, doi: 10.1029/2009GL039035.
ANTARCTIC WARMING
Two decades of temperature change in Antarctica. NASA Earth Observatory. 2007 Nov 21. http://earthobservatory.nasa.gov
Wilkins Ice Bridge
Wilkins Ice Bridge collapse. 2009 April 8. NASA Earth Observatory. http://earthobservatory.nasa.gov
West Antarctic ice sheet
Melting ice and sea-level rise
Floral responses
Turner J et al. Antarctic climate change and the environment. The Scientific Committee on Antarctic Research. Scott Polar Research Institute, Cambridge, UK; 2009.
ARCTIC
Arctic change. Ice – sea ice. www.arctic.noaa. gov/detect/ice-seaice.shtml [Accessed 2011 March 15].
Potential shipping routes. European Space Agency www.esa.int [Accessed 23 Jan 2010].
"What direction are we taking…"
Doyle A. Arctic melt threatens indigenous people. Reuters. 2007 Oct 2. www.reuters.com
GREENLAND MELT
Steffen K et al. The melt anomaly of 2002 on the Greenland Ice Sheet from active and passive microwave satellite observations. J Geophys Res Lett 2004;31(20).
Hanna H et al. Runoff and mass balance of the Greenland ice sheet: 1958-2003. J Geophys Res 2005;110: D13108, doi:10.1029/2004JD005641.
Greenland Melt Extent 2007. Konrad Steffen and Russell Huff Cooperative Institute for Research in Environmental Sciences (CIRES), University of Colorado at Boulder, CO 80309-0216 http://cires.colorado.edu

3 Shrinking Glaciers
GLACIAL RETREAT
World Glacier Monitoring Service. www.geo.unizh.ch/wgms/mbb/sum08.html
United Nations Environment Programme. 2009. Global glacier changes: facts and figures. www.grid.unep.ch/glaciers/
World Global Monitoring Service. Fluctuations of glaciers (FoG) 1995–2000: www.geo.unizh.ch/wgms/fog.html
WWF, Going, going, gone. Climate change and global glacier decline: www.panda.org
Mastny L, Worldwatch Institute. Melting of Earth's ice cover reaches new high. www.upe.ac.za
Dyurgerov MB, Meier MF. Glaciers and the changing Earth system: A 2004 snapshot. Institute of Arctic and Alpine Research. Occasional Paper 58; 2005.
THINNING OVER TIME
THINNING
World Glacier Monitoring Service op. cit.
United Nations Environment Programme.

2009. Global glacier changes: facts and figures. www.grid.unep.ch/glaciers/

USA
US Geological Survey monitoring of South Cascade glacier http://ak.water.usgs.gov
Tankersley J. South Glacier has shrunk by half since 1958. The Seattle Times. 2009 Aug 6. http://seattletimes.nwsource.com

Northern Andes
Peru's Quelccaya glacier could disappear within 10 years, specialist says. Andean Air Mall & Peruvian Times. 2010 Aug 6. www.peruviantimes.com

European Alps
Alean J. SwissEduc www.swisseduc.ch www.glaciers-online.net

Tien Shan
Aizen VB, Aizen FM, Kuzmichonok VA. Glaciers and hydrological changes in the Tien Shan: simulation and prediction. Environmental Research Letters 2007;2(4).

Irian Jaya
Kincaid JL, Klein AG. Retreat of the Irian Jaya glaciers from 2000 to 2002 as measured from IKONOS satellite images. 61st Eastern Snow Conference, Portland, Maine, USA; 2004. www.easternsnow.org

Glacial lakes
www.visibleearth.nasa.gov

4 Ocean Changes
Bindoff NL et al. Observations: oceanic climate change and sea level In: Solomon S et al. editors. Climate change 2007: The physical science basis. Contribution of Working Group I to the Fourth Assessment Report of the Intergovernmental Panel on Climate Change . Cambridge, UK and New York, NY, USA: Cambridge University Press; 2007.
Hoegh-Guldberg O, Bruno JF. The impact of climate change on the world's marine ecosystems. Science 2010;328:1523-8.
Meehl GA et al. Global climate projections. In: Solomon S, 2007 op. cit.
Orr JC et al. Research priorities for ocean acidification. Report from the Second Symposium on the Ocean in a High-CO_2 World. Monaco: SCOR, UNESCO-IOC, IAEA, and IGBP 2008 Oct 6-9.
Cooley SR, Doney SC. Anticipating ocean acidification's economic consequences for commercial fisheries. Environ Res Letters 2009;4:1-8.

Coral reef bleaching
Hoegh-Guldberg O et al. Coral reefs under rapid climate change and ocean acidification. Science 2007 Dec;318(5857):1737-42.

OCEAN SURFACE TEMPERATURE ANOMALIES
National Climate Data Center. The Annual Global Ocean Temperature Anomalies (degrees C) 2010 ftp.ncdc.noaa.gov [Accessed 2010 21 June].

OCEAN ACIDITY
Sabine CL et al. The oceanic sink for anthropogenic CO_2. Science 2004;305(5682), 367-71.

IMPACT OF ACIDITY
Ocean acidification. www.antarctica.gov.au
Photos: J Cubillos

5 Everyday Extremes
Slingo J. Pakistan floods and extreme weather in August 2010. UK Met Office. www.metoffice.gov.uk

Knutson TR et al. Tropical cyclones and climate change. Nature Geoscience 2010:3;157-163.
CBC News. China floods. Landslides leave 1,148 missing. CBCnews. 2010 Aug 9. www.cbc.ca
Trenberth KE et al. Observations: Surface and atmospheric climate change. In: Solomon S et al. editors. Climate change 2007: the physical science basis. Contribution of Working Group I to the Fourth Assessment Report of the Intergovernmental Panel on Climate Change . Cambridge, UK and New York, NY, USA: Cambridge University Press; 2007.

73% of land area…
Alexander LV et al. 2006. Note that this applies primarily to the northern hemisphere as there is a lack of sufficient records in the tropics and southern hemisphere.

INCREASE IN WARM NIGHT
CHANGE IN LENGTH OF WARM SPELLS
TREND IN INTENSITY OF RAINFALL
Alexander LV et al. Global observed changes in daily climate extremes of temperature and precipitation. J Geophys Res 2006;111: D05109.

HEAT WAVE IN EUROPE
Stöckli R, Simmon R, Herring D. NASA Earth Observatory, based on data from the MODIS land team.
Garcia-Herrera R et al. A review of the European summer heat wave of 2003. Critical Reviews in Environmental Science and Technology. 2010;40(4):267-306.

Part 2: The Changing Climate
"Despite the large uncertainties…"
Schneider, S. 2009. Science as a contact sport. National Geographic: Washington, DC. http://stephenschneider.stanford.edu/SAACS/saacs_book.htm Steve, a leading climate scientist, died on 19 July 2010 at the age of 65.

6 The Greenhouse Effect
Forster P et al. Changes in atmospheric constituents and in radiative forcing. In: Solomon S et al. editors. Climate change 2007: the physical science basis. Contribution of Working Group I to the Fourth Assessment Report of the Intergovernmental Panel on Climate Change. Cambridge, UK and New York, NY, USA: Cambridge University Press; 2007: 129-234

ATMOSPHERIC CHANGE
ATMOSPHERIC LIFETIME
APPROXIMATE CONTRIBUTION OF GHGS
Blasing TJ. Recent greenhouse gas concentrations, DOI: 10.3334/CDIAC/atg.032. Updated Dec. 2009. Carbon Dioxide Information Analysis Center http://cdiac.ornl.gov[Accessed 2010 Jan 13].

7 The Climate System
The impacts of a warming Arctic: Artic Climate Impact Assessment. Cambridge: Cambridge University Press; 2004.
Alley RB et al. Abrupt climate change. Science 2002;299 (5615):2005-10.
Broecker WS, Thermohaline circulation, the Achilles heel of our climate system: will man-made CO_2 upset the current balance? Science 1997;278:1582–88.
Fedorov AV, Philander SG. Is El Niño changing? Science 2000;288:1997–2002.
Stocker TF, Marchal O. Abrupt climate change

in the computer: is it real? Proceedings of the National Academy of Sciences of the United States of America (PNAS) 2000;97:1362–65.
Sutton RT, Hodson DLR, Atlantic ocean forcing of North American and European summer climate. Science 2005;309:115–18.
Trenberth K. Uncertainty in hurricanes and global warming. Science 2005;380 (5729):1753-54.

"The climate system…"
IPCC. Annex B. Glossary of Terms. www.ipcc.ch

Tropics
Christensen JH et al. Regional climate projections. In: Solomon S et al. editors. Climate change 2007: the physical science basis. Contribution of Working Group I to the Fourth Assessment Report of the Intergovernmental Panel on Climate Change. Cambridge, UK and New York, NY, USA: Cambridge University Press; 2007.

Jet stream
Kuhlbrodt T et al. An integrated assessment of changes in the thermohaline circulation. Climatic Change 2009;96(4):489-537.
Lozier M Susan. Deconstructing the conveyor belt. Science 2010;328:1507-11.
Meehl GA et al. Global climate projections. In: Climate Change 2007. op. cit.

Storms
Meehl GA et al. Global climate projections. In: Climate Change 2007. op. cit.

Northern Europe
Bischof B, Mariano AJ, Ryan EH. The North Atlantic drift current 2003 [cited June 3 2010.. http://oceancurrents.rsmas.miami.edu/atlantic/north-atlantic-drift.html.
Kuhlbrodt T 2009 op. cit.

Europe
Della-Marta PM et al. Summer heat waves over western Europe 1880-2003, their relationship to large-scale forcings and predictability. Climate Dynamics 2007;29 (2-3):251-75.

Sahel
Shanahan T M et al. Atlantic forcing of persistent drought in West Africa. Science 2009;324(5925):377-80.

El Niño
Houghton J. Global warming: the complete briefing. 4th ed. Cambridge: Cambridge University Press; 2009.

8 Interpreting Past Climates
Hegerl GC et al. Understanding and attributing climate change In: Solomon S et al. editors. Climate change 2007: the physical science basis. Contribution of Working Group I to the Fourth Assessment Report of the Intergovernmental Panel on Climate Change. Cambridge, UK and New York, NY, USA: Cambridge University Press; 2007.
Houghton J. Global warming: The complete briefing. 4th ed. Cambridge: Cambridge University Press; 2009.
Lüthi D et al. High-resolution carbon dioxide concentration record 650,000-800,000 years before present. Nature 2008;453(7193):379-82.
National Research Council. Surface temperature reconstructions for the last 2,000 years. Washington DC: National Academies Press; 2006.
Tans P. Recent global CO_2. 2010. www.esrl.noaa.gov [Accessed 2010 June 23].

ACCUMULATED KNOWLEDGE
Web of Science – Science and social science citation indices. www.isiknowledge.com

CO₂ FLUCTUATIONS
LINK BETWEEN CO₂ AND TEMPERATURE
IPCC www.ipcc.ch/present/graphics.htm
SURFACE TEMPERATURES
Graphic based on NRC. Surface temperature reconstructions for the last 2,000 Years. Washington, DC: National Academies Press; 2006. Multiproxy set 1: Mann & Jones 2003; multiproxy set 2: Moberg et al. 2005; tree rings: Espert et al. 2002; multiproxy set 3: Hegerl et al. 2006; borehole temperatures: Huang et al. 2000; glacier lengths: Oerlemans 2005b.
ACCOUNTING FOR WARMING
Hegerl GC et al 2007 op. cit.

9 Forecasting Future Climates
International Panel on Climate Change (IPCC) Fourth Assessment Report, Working Group I, The Physical Science Basis. Summary for Policymakers. IPCC; 2007. www.ipcc.ch
K Ruosteenoja et al, Future climate in world regions: an inter-comparison of model-based projections for the new IPCC emissions scenarios. Finnish Meteorological Institute, Helsinki; 2003. www.ipcc-data.org
Hadley Centre, Uncertainty, risk and dangerous climate change; recent research on climate change science. Hadley Centre, Exeter; 2004
Stainforth DA et al. Letter: Uncertainty in predictions of the climate response to rising levels of greenhouse gases. Nature 2005;433:403-06 www.nature.com
Nakicenovic N, Swart R. editors. Special report on emissions scenarios. IPCC and Cambridge University Press; 1996. www.grida.no
SCIENTIFIC CONFIDENCE IN PREDICTIONS OF CLIMATE CHANGE
IPCC, 2007, op cit. The IPCC adopted a formal definition of uncertainties: Virtually certain: probability of being true is greater than 99%; Extremely likely: greater than 95%; Very likely: greater than 90%; Likely: greater than 66%.
STABILIZATION SCENARIOS
IPCC 2007 op. cit. Figure 5.1. www.ipcc.ch
EMISSION SCENARIOS
PROJECTED WARMING
IPCC 2007 op. cit. Figure SPM.5 www.ipcc.ch
LOCAL WARMING
PRECIPITATION
Stainforth D, personal communication. www.ClimatePrediction.net

10 Tipping Elements
Rockström, J et al. Planetary boundaries: exploring the safe operating space for humanity. Ecology and Society 2009;14(2):32. www.ecologyandsociety.org
Lenton TM et al. Tipping elements in the Earth's climate system. PNAS 2008;105(6):1786-93.
Shellnhuber HJ. Tipping elements in the Earth System. PNAS 2009;106 (49):20561-63.
IOM. Compendium of IOM's activities in migration, climate change and the environment. Geneva: International Organization for Migration; 2009.
IOM. Migration, environment and climate change: assessing the evidence. Geneva: International Organization for Migration; 2009.
Barnett J, Webber M. Accommodating migration to promote adaptation to climate change. Background document for the Commission on Climate and Development. Melbourne: Department of Resource Management and Geography, The University of Melbourne; 2009. www.ccdcommission.org
Pew Center on Global Climate Change. The science and consequences of ocean acidification. Science Brief 3. Arlington: Pew Center; 2009.

The global burden of migration…
Estimates of the number of 'climate migrants' by 2050 vary widely, with extreme estimates at 1 billion. There is little empirical evidence to draw upon, and we prefer to be cautious. 100 million by 2050 is half of the 'conventional wisdom' at the moment. An additional number would be displaced on a temporary basis – moving back to their homelands after a flood for instance.
IOM. Migration, environment and climate change: assessing the evidence. op. cit.
FRAGILE STATES AND CLIMATE CRISES
Smith D, Vivekananda J. A climate of conflict: The links between climate change, peace and war. International Alert: London; 2007. www.InternationalAlert.org
Relocating vulnerable communities
Interview with Elizabeth Marino, an anthropologist who has worked on coastal migration in Alaska for 10 years. She is affiliated with the University of Alaska and features in a forthcoming documentary on climate change and migration.
Marino E. Imminent threats, impossible moves, and unlikely prestige: Understanding the struggle for local control as a means towards sustainability. The United Nation University Journal on Environment and Human Security 2009;12:42-50.
Marino E, Schweitzer P. Talking and not talking about climate change. In: Anthropology and climate change. Crate S, Nuttall M, editors. Walnut Creek, CA: Left Coast Press; 2008. pp. 209-217.
Schweitzer, P, Marino E. Shishmaref Co-location Cultural Impact Assessment. Seattle, Washington: Tetra Tech Inc; 2005.
Harwood S, Carson D, Marino E. In press. Weather hazards, Place and resilience in the Remote North. In Carson D et al, editors. Demography at the edge: Remote human populations in developed nations. Farnham, UK: Ashgate Publishing.
Drought and complex emergencies
Interview with Mohamed Hamza, a global expert on disasters and humanitarian crises. He leads the Adaptation Academy developed by the Global Climate Adaptation Partnership. See the Climate Change, Environment and Migration Alliance for additional material: www.ccema-portal.org
Black R et al. Demographics and climate change: Future trends and their policy implications for migration. Working Paper T-27, Development Research Centre on Migration, Globalisation and Poverty. Sussex: University of Sussex; 2008. www.migrationdrc.org
UNEP programme on conflicts and disasters: www.unep.org
Coastal insecurity
Interview with Sujatha Byravan (Centre for Development Finance) and Chella Rajan (Indian Institute of Technology, Madras). They work on humanitarian crises in coastal regions and rights-based approaches to climate adaptation.
Byravan S, Rajan SC. Providing new homes for climate change exiles. Climate Policy 2006;6(2).
Byravan S, Rajan SC. Social impacts of climate change. South Asia Journal of Migration and Refugee Issues 2009;5(3):134.

Part 3: Driving Climate Change
"This is not fiction…"
Obama's speech to the Copenhagen Climate summit. www.guardian.co.uk

11 Emissions Past & Present
SHARE OF ANNUAL EMISSIONS
CUMULATIVE CARBON EMISSIONS
SOURCE OF GHGS
COMPARATIVE GROWTH
Climate Analysis Indicators Tool (CAIT) Version 7.0. Washington, DC: World Resources Institute; 2010. http://cait.wri.org [Accessed 2011 May 24].
CO₂ IN THE ATMOSPHERE
UNEP Vital Graphics, quoting David J Hofmann of the Office of Atmospheric Research at the National Oceanic and Atmospheric Administration, March 2006: www.grida.no/climate
Blasing TJ. Recent greenhouse gas concentrations. DOI: 10.3334/CDIAC/atg.032. Updated Dec. 2009. Carbon Dioxide Information Analysis Center. http://cdiac.ornl.gov [Accessed 2011 March 29].
Neftel A et al. Historical carbon dioxide record from the Siple Station ice core. In Boden TA et al, editors. Trends'93: A compendium of data on global change. ORNL/CDIAC-65. Carbon Dioxide Information Analysis Center. Oak Ridge National Laboratory; 1994.
Keeling CD, Whorf TP. Carbon Dioxide Research Group, Scripps Institution of Oceanography, University of California; 2001.

12 Fossil Fuels
IEA Statistics: CO₂ emissions by sector. Total CO₂ emissions from fuel combustion. 2010 highlights. International Energy Agency. Paris, France. www.iea.org
Boden TA, Marland G, Andres RJ. Global, regional, and national fossil-fuel CO₂ emissions. Carbon Dioxide Information Analysis Center. US Department of Energy; 2009. doi 10.3334/CDIAC/00001 http://cdiac.ornl.gov/trends/emis/tre_prc.html. [Accessed 2010 Feb 10].
International Energy Agency, IEA. Electricity/Heat Data for China, People's Republic of http://iea.org/stats [Accessed 2010 March 14].
Le Quere C, Raupach MR, Canadell JG et al. Trends in the sources and sinks of carbon dioxide. Nature Geoscience 2009;2:831-836.
CARBON DIOXIDE EMISSIONS GROWTH
FOSSIL FUEL BURNING
EMISSIONS PER PERSON
FUEL SOURCE
International Energy Agency. CO₂ emissions from fuel combustion. 2010 highlights. www.iea.org [Accessed 2011 March 29].
COAL RESERVES
World Energy Council. 2010 Survey of energy resources. Table 1.1. www.worldenergy.org [Accessed 2011 June 3].

13 Methane & Other Gases
METHANE
NITROUS OXIDE
Climate Analysis Indicators Tool (CAIT) Version

7.0. Washington, DC: World Resources Institute, 2010. http://cait.wri.org [Accessed March 2010].

Methane hydrates
Goel N. In situ methane hydrate dissociation with carbon dioxide sequestration: current knowledge and issues. Journal of Petroleum Science and Engineering; 2006 www.elsevier.com

Glasby GP. Potential impact on climate of the exploitation of methane hydrate deposits offshore. Marine and Petroleum Geology 2003;20:163-75.

Kono H et al. Synthesis of methane gas hydrate in porous sediments and its dissociation by depressurizing. Powder Technology 2002;122:239–46.

Shakhova N et al. Extensive methane venting to the atmosphere from sediments of the East Siberian Arctic Shelf. Science 2010;327:1246-50.

COMPARATIVE GHGS EMISSIONS
Climate Analysis Indicators Tool (CAIT) Version 7.0. (Washington, DC: World Resources Institute, 2010). Compare Gases. [Accessed 2010 Mar 18].

COMPARATIVE GLOBAL WARMING POTENTIAL
Forster P et al. Changes in atmospheric constituents and in radiative forcing. In: Solomon S et al. editors. Climate change 2007: the physical science basis. Contribution of Working Group I to the Fourth Assessment Report of the Intergovernmental Panel on Climate Change. Cambridge, UK and New York NY, USA: Cambridge University Press; 2007.

14 Transport
International Road Federation. World Road Statistics 2009. Geneva: International Road Federation; 2009.

US Energy Information Administration. Emissions of greenhouse gases in the United States 2008. DOE/EIA-0573(2008), Washington DC: US Energy Information Administration; 2009.

Annual vehicle-miles of travel, 1980-2008 by functional system, national summary. In: Highway Statistics 2008. US Federal Highway Administration; 2009.

TRANSPORT EMISSIONS
IEA Statistics: CO_2 emissions by sector. Total CO_2 emissions from fuel combustion, road. 2010 highlights. International Energy Agency. Paris, France. www.iea.org

INTERNATIONAL SHIPPING AND AVIATION INCREASING EMISSIONS
International Energy Agency. CO_2 Emissions from Fuel Consumption: Highlights. Paris: IEA; 2009.

CARS
International Road Federation. World Road Statistics 2009. Geneva: International Road Federation; 2009.

15 Agriculture
McIntyre BD et al, editors. International assessment of agricultural knowledge, science and technology for development. Agriculture at a crossroads: Global report. Washington: Island Press; 2009.

WRI EarthTrends, accessed 2 April 2011, www.EarthTrends.wri.org

"It is absolutely immoral…"
Brabeck-Letmathe P. Personal communication.

2011 April 3. See also: A conversation with Peter Brabeck-Letmathe. Independent on Sunday. Article on his presentation to the Council on Foreign Relations (CFR) in New York. 2011 Mar 23.

AGRICULTURAL GHG EMISSION
CAIT 2007 [Accessed 2010 March 21].
World Development Indicators [Accessed 2010 March 21].

COMPARATIVE PROFILES
Hoffman U. Assuring food security in developing countries under the challenges of climate change: Key trade and development issues of a fundamental transformation of agriculture. Geneva: UNCTAD; 2011.

COMPARATIVE EMISSIONS
Kasterine A. Counting carbon in exports: Carbon footprinting initiatives and what they mean for exporters in developing countries. In: ITC. International Trade Forum 2010;1:30-31. www.intracen.org

Keane J et al. Climate change and developing country agriculture: An overview of expected impacts, adaptation and mitigation challenges, and funding requirements. ICTSD/ International Food & Agricultural Trade Policy Council. Issue Brief no. 20; 2009. www.ictsd.org

Edwards-Jones G et al. Variability of exporting nations to the development of a carbon label in the United Kingdom. Environmental Science & Policy 2009;12:479-90.

WORLD AGRICULTURAL EMISSIONS
A comprehensive inventory of agricultural contributions to greenhouse gas emissions has not been produced, due to the difficulty of estimating all of the land-use changes, trade, wastage in consumption, and other factors. Greenpeace have put together the most competent review, based on IPCC and peer-reviewed reports. UNCTAD draws upon the Greenpeace report. The range of GHG estimates by source is 8,500 to 16,500 million tonnes CO_2e: the chart uses the average of the range of values cited in the literature. It is likely an underestimate of total emissions but a reasonable division according to different sources.

Bellarby J et al. 2008. op. cit. Hoffmann U. 2011. Assuring food security in developing countries under the challenges of climate change: Key trade and development issues of a fundamental transformation of agriculture. Geneva: United Nations Conference on Trade and Development. www.unctad.org

16 The Carbon Balance
Houghton RA. Emissions (and sinks) of carbon from land-use change. Woods Hole Research Center. http://cait.wri.org/downloads/ DN-LUCF.pdf

CO_2. The human perturbations. Paris: UNESCO, SCOPE, UNEP. Policy Briefs 10. 2009 Nov. http://unesdoc.unesco.org

By 2030 we will need…
WWF. Living planet report, 2010. Gland: WWF International; 2010.

Ewing B et al. Calculation methodology for the National Footprint Accounts, 2010 edition. Oakland: Global Footprint Network; 2010. www.footprintnetwork.org

CARBON CYCLE
The Carbon Project. www.globalcarbonproject.org
Friedlingstein P, Houghton RA et al. Update on CO_2 emissions. Nature Geoscience

Advanced online publication. 2010 Nov 21.
Le Quéré et al. Trends in the sources and sinks of carbon dioxide. Nature Geoscience 2009;2:831-36.

LAND-USE CHANGE
Net flux of carbon to the atmosphere from land-use change, 2000. http://earthtrends.wri.org

TRENDS IN CARBON SOURCES AND SINKS
www.globalcarbonproject.org

Part 4: Expected Consequences
"The security of people and nations…"
Burke T. The ENDS Report. 2008 May 23.

17 Disrupted Ecosystems
Fischlin A et al. Ecosystems, their properties, goods, and services. In: Parry ML et al. editors. Climate Change 2007: Impacts, adaptation and vulnerability. Contribution of Working Group II to the Fourth Assessment Report of the Intergovernmental Panel on Climate Change. Cambridge University Press: Cambridge; 2007. pp. 211-72.

Hansen MC et al. Humid tropical forest clearing from 2000 to 2005 quantified by using multitemporal and multiresolution remotely sensed data. Proceedings of the National Academy of Sciences of the United States of America. 2008;105(27):9439-44.

1.5°C to 2.5°C…
International Panel on Climate Change, Working Group II Contribution to the Intergovernmental Panel on Climate Change Fourth Assessment Report. Climate Change 2007: Impacts, adaptation and vulnerability. Summary for policymakers. IPCC: 2007. www.ipcc.ch

"Clearly we are…"
Morales A. Global warming very likely caused by humans, UN says (Update5). 2007 Feb 2. www.bloomberg.com

Non-native species
Fischlin A et al. 2007. op. cit.

Prairie wetlands
Johnson WC et al. Prairie wetland complexes as landscape functional units in a changing climate. Bioscience 2010;60(2):128-40.

Brazilian Cerrado
Thomas CD et al. Extinction risk from climate change. Nature 2004;427:145-48.

Adélie Penguins
Ducklow HW et al. Marine pelagic ecosystems: The West Antarctic Peninsula. Philosophical Transactions of the Royal Society B-Biological Sciences 2007;362(1477):67-94.

Changes consistent with climate change
Rosenzweig C et al. Attributing physical and biological impacts to anthropogenic climate change. Nature 2008;453:353-U320.

Europe
Climate change threatens new wave of extinction. 2004 Jan 9. Based on Nature 2004;427;145-48. www.birdlife.org

Kirby A. Climate risk 'to million species'. 2004 Jan 7. www.news.bbc.co.uk

Sundarbans Delta
Loucks C et al. Sea-level rise and tigers: Predicted impacts to Bangladesh's Sundarbans mangroves, Climatic Change 2010;98(1-2):291-98.

Mount Kilimanjaro
Hemp A. Climate change-driven forest fires marginalize the impact of ice cap wasting on

Kilimanjaro. Global Change Biology 2005;11:1013-23.

Australia
Yates CJ et al. Assessing the impacts of climate change and land transformation on Banksia in the South West Australian Floristic region. Diversity and Distributions 2010;16 (1):187-201.

Ocean warming
Van Bressem MF. Emerging infectious diseases in cetaceans worldwide and the possible role of environmental stressors. Diseases of Aquatic Organisms 2009;86 (2):143-57.

GLOBAL WARMING
Leemans R, Eickhout B. Another reason for concern: regional and global impacts on ecosystems for different levels of climate change. Global Environment Change 2004;14:219–28. www.elsevier.com

18 Water Security

Bates BC et al, editors. Climate change and water. Technical paper of the Intergovernmental Panel on Climate Change. Geneva: IPCC Secretariat; 2008. pp 210.
World Bank. Economics of adaptation to climate change: Synthesis report. Washington: World Bank; 2010. http://climatechange.worldbank.org
UNEP. Vital water. www.unep.org
Alcamo J, Henrichs T. Critical regions: A model-based estimation of world water resources sensitive to global changes. Water Policy 2002;64: 352–62.
Alcamo J, Henrichs T, Rösch T. World water in 2025: global modeling and scenario analysis for the world commission on water for the 21st century. Kassel World Water Series Report 2. Center for Environmental Systems Research, University of Kassel, Germany; 2000.
Arnell NW. Climate change and global water resources: SRES emissions and socio-economic scenarios. Global Environmental Change 2004;14: 31–52.
UNESCO Water e-Newsletter No.123: Mountains. 2005 Dec 5. www.unesco.org [Accessed 2005 Dec 11]. Based on work of Lisa Mastny, Worldwatch Institute news brief 00–02.

Nearly 3 billion people…
Molden, D. (ed.) Water for food, water for life: A comprehensive assessment of water management in agriculture. London: Earthscan, and Colombo: International Water Management Institute (2007).

Up to 5 billion people…
Lucinda Mileham. 15 September 2010. Water security and climate change: Facts and figures. www.scidev.org
FRESHWATER RESOURCES
FAO Aquastat. www.fao.org [Accessed 2011 May 1]
TREND IN RIVER RUN-OFF
Dai A. Drought under global warming: a review. WIREs Clim Change 2011;2:45–65. Doi: 10.1002/wcc.81.
Trenberth KE. Changes in precipitation with climate change. Climate Research 2011;47:123-138.
Potential change in run-off
US Geological Survey. Cited in: Bates BC et al; 2008 op. cit.
This map is based on the average results from 12 global climate models from the IPCC fourth assessment report. It shows some of the areas where climate impacts are

uncertain. However, confidence in local predictions of future climate change impacts on water resources is generally low.

19 Food Security

Lobelle DB, Schlenker W, Costa-Roberts J. Climate trends and global crop production since 1980. DOI: 10.1126/science.1204531.
McIntyre BD, Herren HR, Wakhungu J, Watson RT, editors. International assessment of agricultural knowledge, science and technology for development: Global report. Washington DC: Island Press; 2009.
Lobell DB et al. Prioritizing climate change adaptation needs for food security in 2030. Policy Brief. Stanford: Stanford University Program on Food Security and the Environment; 2008.
Lobell DB et al. Prioritizing climate change adaptation needs for food security in 2030. Science 2008; 319: 607-610.
Tubiello FN, Fisher G. Reducing climate change impacts on agriculture: global and regional effects of mitigation 2000–2080. ILASA; 2006 (draft paper).
Fischer G, Shah M, van Velhuizen H. Climate change and agricultural vulnerability (working paper) Laxenbourg: IIASA; 2002. Based on Hadley Centre model H adCM3.
Fischer G et al. Socio-economic and climate change impacts on agriculture: an integrated assessment 1990–2080, Philosophical Transactions of the Royal Society B: Biological Sciences, 2005 (published online). www.pubs.royalsoc.ac.uk
Investment of $7 billion a year…
MALNOURISHED CHILDREN
Nelson G et al. Climate change: Impact on agriculture and costs of adaptation. Washington: International Food Policy Research Institute (IFPRI); 2010.
McIntyre BD et al. 2009. op cit
ESCALATING FOOD PRICES
News stories on rising food prices and their role in triggering protests have been common in the media. However, the linkage to political unrest is not a simple climate-impact causal relationship.
FAO. www.fao.org/worldfoodsituation/wfs-home/foodpricesindex/en/
Johnstone S, Mazo J. Global warming and the Arab Spring. Survival 2011;53:11-17.
Breisinger C et al. Food security and economic development in the Middle East and North Africa. International Food Policy Research Institute discussion paper 00985; 2010. www.ifpri.org
IMPACT OF CLIMATE CHANGE ON CROP PRODUCTION
Lobell DB et al. Science. 2008. op. cit.
Lobell DB et al. Stanford Policy Brief; 2008. op. cit.

20 Threats to Health

Akhatar R et al, Human health. In: McCarthy JJ et al, Climate Change 2001: Impacts, Adaptation, and Vulnerability. Cambridge: Cambridge University Press; 2001. pp 451–86
Epstein P, Mills E, editors. Climate change futures: Health, ecological, and economic dimensions. Cambridge MA: Center for Health and the Global Environment and Harvard Medical School; 2005.
Brownstein J. Lyme disease: implications of climate change. In: Epstein P, Mills E, editors. 2005. op. cit. pp. 45–47.

Ebi KL et al. Infectious and respiratory diseases: malaria. In: Epstein P, Mills E. op cit, pp. 32–41.
Kutz SJ et al. Emerging parasitic infections in arctic ungulates. Integrative and Comparative Biology, April 2004;44 (2):109–18.
World Health Organization (WHO), World Malaria Report 2005, WHO, Geneva. www.rbm.who.int
Global Burden of Disease (GBD). Geneva: WHO. 2010. www.who.int
POTENTIAL FUTURE MALARIA RISK
European Environment Agency. The European Environment. Luxembourg: EEA; 2010.
Rogers DJ, Randolph S. The global spread of malaria in a future, warmer world. Science 8 September 2000;289(5485): 1763-1766.
INCREASED DISEASE BURDEN
Ebi KL Adaptation costs for climate change-related cases of diarrhoeal disease, malnutrition, and malaria in 2030. Globalization and Health 2008;4:9 doi:10.1186/1744-8603-4-9.
Projected excess incident cases of climate-related diseases (000s) for Unmitigated scenario (UE), high estimate, 2030; scenario approximately follows the Intergovernmental Panel on Climate Change IS92a or business as usual scenario.
WHO. The World Health Report 2002. Geneva: WHO; 2002.
Murray DJL. Quantifying the burden of disease – the technical basis for disability-adjusted life years. Bulletin of the World Health Organization 1994;72(3): 429-445.
McMichael AJ et al. Climate change. In: Comparative quantification of health risks. Geneva: WHO; 2003.
Tick-borne diseases
Brownstein J, op cit.

21 Rising Sea Levels

Vermeera M, Rahmstorf S. Global sea level linked to global temperature. Proceedings of the National Academy of Science; 2009. 106(51): 21527-21532. www.pnas.org
Rahmstorf S. A semi-empirical approach to projecting future sea-level rise. Science 2007;315(5810):368-370.
A semi-empirical relation is presented that connects global sea-level rise to global mean surface temperature. It is proposed that, for time scales relevant to anthropogenic warming, the rate of sea-level rise is roughly proportional to the magnitude of warming above the temperatures of the pre–Industrial Age. This holds to good approximation for temperature and sea-level changes during the 20th century, with a proportionality constant of 3.4 millimeters/year per °C. When applied to future warming scenarios of the Intergovernmental Panel on Climate Change, this relationship results in a projected sea-level rise in 2100 of 0.5 to 1.4 meters above the 1990 level.
Bindoff NL et al. Observations: oceanic climate change and sea level. In: Solomon S et al. editors. Climate change 2007: the physical science basis. Contribution of Working Group I to the Fourth Assessment Report of the Intergovernmental Panel on Climate Change. Cambridge, UK and New York, NY, USA: Cambridge University Press; 2007. www.ipcc.ch
Meehl GA et al. Global climate projections. In: Solomon S et al op cit. 2007.

WETLANDS LOSS
EFFECT OF SEA-LEVEL RISE
DIVA runs by the University of Southampton for a medium scenario of sea-level rise: A1B medium regionalized without adaptation, number of people flooded by country. Courtesy of Robert Nicholls and Sally Brown.
Disappearing islands
Kirby A. Islands disappear under rising seas. 1999 June 14. news.bbc.co.uk
Evacuated islands
Robinson E et al. Dwindling coastlines – Hazards of the Jamaican coastline. Jamaica Gleaner. 2006 Jan 27. www.jamaica-gleaner.com
Abandoned island
Climate change and sea-level ris. www.sidsnet.org
Threatened island
Move Tuvalu population to a Fiji island to ensure survival, scientist says. 2006 Feb 20. www.tuvaluislands.com
NILE DELIA
UNEP vital climate graphics: www.grida.no
Under a 1-meter sea-level rise...
Agrawala S et al. Development and climate change. In: Egypt: Focus on Coastal Resources and the Nile. OECD; 2004.
Dasgupta, S et al. The impact of sea level rise on developing countries: a comparative analysis. World Bank Policy Research Working Paper 4136. 2007. www.worldbank.org
PROJECTED SEA-LEVEL RISE
The shaded areas are based on four central estimates for the A1B medium regionalized scenario. The top two lines are for more extreme scenarios that include ice melt from glaciers and ice caps (lower line: Rahmstorf 2007 and upper line: Vermeera and Rahmstorf 2009). All are from DIVA run by the University of Southampton. Courtesy of Robert Nicholls and Sally Brown.

22 Cities at Risk
Anthoff D et al. Global and regional exposure to large rises in sea-level: a sensitivity analysis: Tyndall Centre Working Paper 96; 2006.
Center for International Earth Science Information Network (CIESIN), CSD Coastal population indicator: data and methodology page. Socioeconomic Data and Applications Center (SEDAC), Columbia University 2006. sedac.ciesin.columbia.edu [Accessed 2006 May 26].
CIESIN. Low Elevation Coastal Zone (LECZ) Urban Rural Estimates, Global Rural-Urban Mapping Project (GRUMP), SEDAC, Columbia University 2010. sedac.ciesin. columbia.edu[Accessed 2010 May 28].
Kamal-Chaoui L, Robert A. editors. Competitive cities and climate change, OECD regional development working papers No 2. Paris: OECD Publishing: 2009.
Port cities
Nicholls RJ et al. Ranking port cities with high exposure and vulnerability to climate extremes. Environment Working Papers No. 1. Paris, France: OECD; 2007.
Almost 22,000...
LIVING ON THE EDGE
CIESIN. LECZ, GRUMP. op. cit.
Population Division of the Department of Economic and Social Affairs of the United Nations Secretariat, World Urbanization

Prospects: the 2009 Revision. www.un.org/esa/population [Accessed 2010 May 31].
New York
Vafeidis A, Flood Hazard Research Centre, University of Middlesex, personal communication
Climate impacts in New York City: sea level rise and coastal floods. NASA. Goddard Institute for Space Studies. icp.giss.nasa.gov
London
London Climate Change Partnership. www.london.gov.uk/lccp/
Marsh TJ. The risk of tidal flooding in London. www.ecn.ac.uk/iccuk/indicators/10.htm
New Orleans
Sigma 2006. Zurich: SwissReinsurance Company. www.swissre.com
Lagos
Ibe AC, Awosika LF. Sea level rise impact on African coastal zones. In: Onide SH, Juma C. editors. A change in the weather: African perspectives on climate change. African Centre for Technology Studies. Nairobi, Kenya; 1991. pp. 105-12. www.ciesin.columbia.edu
Mumbai
Vafeidis A. Flood Hazard Research Centre, University of Middlesex. personal communication

23 Cultural Losses
Case studies of climate change and world heritage. Paris: UNESCO World Heritage Centre; 2007.
Arctic
Arctic Climate Impact Assessment (ACIA). Impacts of a warming Arctic. Cambridge University Press; 2004.
Thoreau's Woods
Willis CG et al. Phylogenetic patterns of species loss in Thoreau's woods are driven by climate change. Proceedings of the National Academy of Sciences of the United States of America 2008;105(44):17029-33.
Northeast USA
Huntington TG et al. Climate and hydrological changes in the northeastern United States: recent trends and implications for forested and aquatic ecosystems. Canadian Journal of Forest Research-Revue Canadienne De Recherche Forestiere 2009;39:199-212.
Scotland, UK
Breeze DJ. Foreword in: Dawson T. editor. Conference proceedings: coastal archaeology and erosion in Scotland. Edinburgh: Historic Scotland; 2003.
Netherlands
Visser H, Petersen AC. The likelihood of holding outdoor skating marathons in the Netherlands as a policy-relevant indicator of climate change. Climatic Change 2009;93:39-54.
Czech Republic
Johnston R et al. Czech Republic's cultural losses mount in flood's wake. 2002: portal. unesco.org
Mt Everest
Legal steps taken to protect natural heritage sites from climate change. 2004 Nov 25. www.iema.net
Alexandria, Egypt
Mostafa MH, Grimal N, Nakashima D, editors. Underwater archaeology and coastal management: focus on Alexandria. Paris: UNESCO; 2000.
El-Raey M. Impacts and adaptation to climate

change on Mediterranean coastal zone of Egypt. University of Alexandria. 2007 May 12. www.planbleu.org
Cherry Blossom, Japan
Miller-Rushing AJ et al. Impact of global warming on a group of related species and their hybrids: Cherry tree (Rosaceae) flowering at Mt. Takao, Japan. American Journal of Botany 2007;94(9):1470-78.
Thailand
Draper M. Floods threaten Thai monuments, Archaeology 1996;49 (2). www.archaeology.org
West Coast National Park, South Africa
Roberts D. South African west coast. Langebaan footprints: a walk with Eve. The Cape Odyssey, 2002 June/July. www.sawestcoast.com
Venice, Italy
Case studies of climate change and world heritage. Paris: UNESCO World Heritage Centre; 2007.
Tuvalu
Farrell B. Pacific islanders face the reality of climate change and of relocation. UN High Commissioner for Refugees News Stories. 2009 Dec 14.

Part 5: Responding to Change
"We are all part of the problem..."
Ban K-M. 2008 April 13. Africa Resource http://www.africaresource.com

24 Urgent Action to Adapt
OCHA. 2010. Pakistan floods emergency response plan revision: 2010 Sept. Geneva: Office for the Coordination of Humanitarian Affairs.
Stern N. The economics of climate change: The Stern review. Cambridge: Cambridge University Press; 2007.
OCHA. Monitoring disaster displacement in the context of climate change: Findings of a study by OCHA and the Internal Displacement Monitoring Centre. OCHA, Geneva; 2009.
Yohe G et al. A synthetic assessment of the global distribution of vulnerability to climate change from the IPCC perspective that reflects exposure and adaptive capacity. New York: CIESIN; 2006. http://sedac.ciesin.columbia.edu
CLIMATIC DISASTERS
EM-DAT: The OFDA/CRED International Disaster Database. Université Catholique de Louvain, Brussels (Belgium) www.cmdat.be [Accessed 2011 May 31].
Sigma. Zurich: SwissReinsurance Company. 2002–10 www.swissre.com
Weather related disasters...
"There is a very human tendency...
Nitin Desai is a member of the Prime Minister's Council on Climate Change, India and Distinguished Fellow, The Energy and Resources Institute (TERI), quoted in: The anatomy of a silent crisis: The human impact report on climate change. Geneva: The Global Humanitarian Forum; 2009. The report documents a range of estimates of the human impact of climate change, and clearly notes the wide uncertainty in the estimates. It concludes that climate change impacts are 300,000 deaths per year and $125 billion in economic damages. Clearly, it is impossible to disentangle the impacts of additional

climate change caused from anthropogenic greenhouse gas emissions from the ongoing impacts of weather and climate. The Atlas authors prefer to report current figures for weather and climate impacts (not solely the additional climate change impacts) and adopt a lower benchmark for our estimates to reflect the wide uncertainty.

The EMDAT data base shows nearly 200,000 deaths from drought, extreme temperature, flood, wet landslides, storms and wildfire for 2006 to 2010. The database does not include deaths from famine and malnutrition as well as the many vector-borne diseases which have a climate component.

Warning for heat waves

A full evaluation of the climate impacts of 2010 has not been compiled in Europe. However, news reports and experts do not anticipate a repeat of the 2003 impacts, in part because the heat wave did not last as long as in 2003.

Pascal M et al. France's heat health watch warning system. Int J Biometeorol 2006 Jan;50(3):144-53. Epub 2005 Nov 23.

Lagadec P. Understanding the French 2003 heat wave experience: Beyond the heat, a multi-layered challenge. Journal of Contingencies and Crisis Management 2004;12(4):160-69. DOI: 10.1111/j.0966-0879.2004.00446.x

More crop per drop

Roberts M. Creating shared value. "More crop per drop". www.creatingsharedvalue.org A Farm Business Advisor (FBA) project, of International Development Enterprises Cambodia. The project won the Creating Shared Value Award sponsored by Nestle in 2010 for its pioneering work as a public-private partnership developing solutions for sustainable development. Photo courtesy of Nestle and IDE-Cambodia.

Lost ecosystems increase climate impacts

In the front line: shoreline protection and other ecosystem services from mangroves and coral reefs. UNEP-WCMC, Cambridge, UK; 2006.

Women's groups manage natural resources The Green Belt Movement. http://greenbeltmovement.org

Tackling the Big Dry

Windram C. Drought in Australia. The lessons we can learn for tackling climate change . Think Carbon; 2009 June. http://thinkcarbon. wordpress.com

Climate variability, climate change and drought in eastern Australia. www.csiro.au [Accessed 2010 Sept].

25 Building Capacity to Adapt

Adaptation learning mechanism. www.adaptationlearning.net

Climate Change Committee www.theccc.org.uk/reports/adaptation

Gigli S, Agrawala S. Stocktaking of progress on integrating adaptation to climate change into development co-operation activities. Paris: OECD; 2007.

Klein RJT et al. Inter-relationships between adaptation and mitigation. In: Parry ML et al, editors. Climate Change 2007: Impacts, adaptation and vulnerability. Contribution of Working Group II to the Fourth Assessment Report of the Intergovernmental Panel on Climate Change. Cambridge University Press: Cambridge, UK; 2007. pp. 745-77.

Nairobi Work Programme

http://unfccc.int/adaptation/items/4159.php

Olhoff A, Schaer C. Screening tools and guidelines to support the mainstreaming of climate change adaptation into development assistance – a stocktaking report. UNDP: New York; 2010.

UK Climate Impacts Programme www.ukcip.org.uk

UNDP. Knowledge management and methodology. www.undp.org/climatechange/ adaptation_knowledge.shtml

Watkiss P, Downing TE, Dyszynski J. Economics of climate adaptation in Africa. Nairobi: UN Environment Programme and Oxford: Stockholm Environment Institute and Global Climate Adaptation Partnership; 2010.

Learn, share, connect worldwide

Learn, share, connect worldwide. www.britishcouncil.org

Niblett H. At the International Youth Forum on Climate Finance. 2010 Sept 9. http://blog.britishcouncil.org

NATIONAL ADAPTATION PLANNING

Adaptation. http://unfccc.int/adaptation/ items/4159.php [Accessed 2011 March].

California

2009 California Adaptation Strategy. California Natural Resources Agency. www.climatechange.ca.gov/adaptation.

Franco G et al. Linking climate change science with policy in California. Climatic Change 2007. DOI 10.1007/s10584-007-9359-8. http://meteora.ucsd.edu

Supporting local adaptation

Rainforest Alliance. www.rainforest-alliance.org Rainforest alliance. uk.ask.com/wiki/ Rainforest_Alliance

Peru

Orlove B. Glacier retreat: Reviewing the limits of human adaptation to climate change. Environment 2009; 51(3)22-34. www.environmentmagazine.org

Huggel C et al. The SDC Climate Change Adaptation Programme in Peru: Disaster risk reduction within an integrative climate change context. In: Ammann WJ et al, editors. Proceedings of the International Disaster and Risk Conference, IDRC, Davos; 2008 August. pp. 276-78. www.forschungsportal.ch

Applying disaster risk reduction for climate change adaptation: country practices and lessons. Geneva: UNISDR. 2009. www.preventionweb.net

Adaptation to climate change by reducing disaster risks: country practices and lessons. Geneva: UNISDR; 2009. www.preventionweb.net

Urban planning

Asian Cities Climate Change Resilience Network. www.rockefellerfoundation.org

African Climate Policy Centre

International Development Research Centre (IDRC) http://publicwebsite.idrc.ca

Africa Adapt. www.africa-adapt.net

Climate Systems Analysis Group. www.csag.uct.ac.za

African Technology Policy Studies Network. www.atpsnet.org

Bangladesh

Documentation from the Asian Development Bank support project: www.adb.org/projects/ project.asp?id=42478

Ministry of Environment and Forests. Bangladesh Climate Change Strategy and Action Plan. Dhaka: MOEF; 2009.

International Centre for Climate Change and Development www.icccad.org

26 City Responses

All websites accessed 2010 August 6–8.

Tokyo Cap-and-Trade program. Tokyo Metropolitan Government; 2009 May 26. www.kankyo.metro.tokyo.jp/en/index.html

European Energy Commission. Covenant of Mayors. www.eumayors.eu

The City Climate Catalogue. ICLEI. Local Governments for Sustainability; 2009. www.climate-catalogue.org

ICLEI. Members. 2010 www.iclei.org

Satterthwaite D. Cities' contribution to global warming: notes on the allocation of greenhouse gas emissions. Environ Urban 2008 Oct;20(2):539-49.

COMMITMENTS

BENEFITS

ICLEI. Mitigation. 2010. www.iclei.org

REDUCTION GOALS

ICLEI. The City Climate Catalogue. op. cit.

C40 CITIES

C40 Secretariat. C40 cities: An introduction. 2010. www.c40cities.org

Chicago

City of Chicago. Chicago Climate Action Plan. 2008 www.chicagoclimateaction.org

London

Collingwood Environmental Planning. The Mayor of London's draft climate change adaptation strategy: Sustainability appraisal report: Non-technical summary. London: Office of the Mayor; 2010 February.

Mexico City

Secretaría del Medio Ambiente Gobierno del Distrito Federal. Mexico City climate action program 2008-12 Summary. 2008. www.sma. df.gob.mx

Lagos

Ministry of Environment Lagos. 2nd Lagos Climate Change Summit Communique. 2010. http://climate-l.org

Johannesburg

Naidoo, R. Plan to cut gas emissions. 2010. www.joburg.org.za

Mumbai

Ghoge K. Climate change action plan for Mumbai in two years. Hindustan Times. 2010 April 1.

Tokyo

Tokyo's urban strategy. Tokyo's big change: The 10-Year Plan. 2010 www.metro.tokyo.jp

Sydney

What the city is doing. 2009 Feb 6. www.cityofsydney.nsw.gov.au

27 Renewable Energy

International Energy Agency. Key World Energy Statistics. Paris: IEA; 2009.

REN21. Renewables Global Status Report: 2010 Update. Paris: REN21 Secretariat; 2011.

International Renewable Energy Agency. www.irena.org [Accessed 2010 Feb 21].

COMMITMENT

REN21. Renewables global status report 2007. Paris: REN21 Secretariat; 2007. Table 2.

REN21. Renewables global status report: 2010 Update. Paris: REN21 Secretariat; 2011. Tables R10, R11.

GROWTH IN WIND ENERGY
RENEWABLE ENERGY SOURCES
RENEWABLE ELECTRICITY PRODUCTION
REN21. Renewables global status report: 2010 Update. Paris: REN21 Secretariat; 2011.

28 Low Carbon Futures
Climate principles
The Climate Group www.theclimategroup.org
Potential reductions
Stockholm Environment Institute. The economics of low carbon, climate-resilient patterns of growth in developing countries: A review of the evidence final report. Submission to DFID. 2010 April. http://sei-international.org
CARBON INTENSITY
CHANGE IN CARBON INTENSITY
IEA Statistics. 2010 highlights. International Energy Agency. Paris, France. www.iea.org

29 Counting Carbon
Eggleston S, Buendia L, Miwa K, Tanabe K, editors. 2006 IPCC Guidelines for National Greenhouse Gas Inventories. Hayama, Japan: Institute for Global Environmental Strategies (IGES); 2006.
United Nations Framework Convention on Climate Change (UNFCCC). Report on the expert meeting on methodological issues relating to reference emission levels and reference levels. In: Subsidiary Body for Scientific and Technological Advice, editor. Geneva: United Nations Office at Geneva; 2009.
US Environmental Protection Agency. Greenhouse Gas Reporting Program. Washington DC: US EPA; 2010. www.epa.gov [Accessed 2010 Nov 25]
REDD+
Gibbs HK, Brown S, Niles JO, Foley JA. Monitoring and estimating tropical forest carbon stocks: making REDD a reality. Environmental Research Letters. 2007;2:1-13.
Rodriguez D. Auditors give boost to REDD project design. 2010 August 18. www.climateactionprogramme.org
CONSUMPTION-BASED ACCOUNTING
CARBON IN TRADED GOODS
Davis SJ, Calderia K. Consumption-based accounting of CO_2 emissions. Proceedings of the National Academy of Science. 2010 March 23;107(12):5678-92. NB: Analysis uses 2004 data.
Around 23%...
Climate Analysis Indicators Tool (CAIT) Version 7.0. Washington, DC: World Resources Institute, 2010.

Part 6: International Policy & Action
www.climateactiontracker.org [Accessed 17 June 2011].
"...Different countries are in..."
Christiana Figueres, Executive Secretary, UN Framework Convention on Climate Change (UNFCCC). Press conference, 2011 May 12. www.un.org

30 International Action
OBSERVERS
UNFCCC application for observer status: http://unfccc.int

SIGNATORIES TO THE UNFCCC AND KYOTO
Secretariat of the UNFCCC: http://unfccc.int [Accessed 2011 April 17]
KEY GROUPINGS WITHIN NEGOTIATIONS
United Nations Framework Convention on Climate Change www.sourcewatch.org
Parties and observers. http://unfccc.int
Party groupings. http://unfccc.int

31 Meeting Kyoto Targets
United Nations Framework Convention on Climate Change (UNFCCC) Kyoto Protocol. [Accessed 2010 June 21] http://unfccc.int/kyoto_protocol/items/2830.php
United Nations Framework Convention on Climate Change (UNFCCC). Copenhagen Accord. 2010 [Accessed 2010 June 21] http://unfccc.int/home/items/5262.php
INCREASED EMISSIONS
Climate Analysis Indicators Tool (CAIT) Version 7.0. Washington, DC: World Resources Institute; 2010.
LONG-TERM EFFECT OF CO_2 EMISSIONS
IPCC. Climate change 2001 summary for policymakers: www.ipcc.ch
PROGRESS TOWARDS KYOTO TARGETS
United Nations Framework Convention on Climate Change. National greenhouse gas inventory for the period 1990–2008. FCCC/SBI/2010/18. Table 6.

32 Looking Beyond Kyoto
NB: The Kyoto Protocol will continue in force beyond 2012, including provisions for the Clean Development Mechanism and Adaptation Fund.
IS IT ENOUGH?
den Elzen MGJ et al. Evaluation of the Copenhagen Accord: Chances and risks for the 2°C climate goal. Bilthoven: Netherlands Environmental Assessment Agency, Ecofys; 2010 May. Report no. 500114017.
WHAT IS ON OFFER UNDER THE COPENHAGEN ACCORD?
Annex I
Secretariat of the United Nations Framework Convention on Climate Change (UNFCCC). Appendix 1. Quantified economy-wide emissions targets for 2020. 2010 12 Aug. http://unfccc.int
Commitments made by largest emitters among developing nations
Secretariat of the UNFCCC. Appendix II. Nationally appropriate mitigation actions of developing country Parties. 2010 August 12. http://unfccc.int
COPENHAGEN ACCORD
CANCUN
Secretariat of the UNFCCC. Copenhagen Accord. 2010 [Accessed 2011 April 9] http://unfccc.int
International Institute for Sustainable Development. A brief analysis of the Copenhagen Climate Change Conference; 2009. www.iisd.org
Secretariat of the UNFCCC. Decision2/CP.15 Copenhagen Accord. 2010 http://unfccc.int
Ad Hoc Working Group on Further Commitments for Annex I Parties under the Kyoto Protocol. Legal considerations relating to a possible gap between the first and subsequent commitment periods. UNFCCC; 2010.

33 Trading Carbon Credits
BUYERS AND SELLERS
Kossoy A, Ambrosi P. State and trends of the carbon market 2010. Washington DC: Carbon Finance at the World Bank; 2010 May.
PROJECT-BASED MARKETS
Donoghue A. UN considers review of alleged carbon offset abuses. Guardian Environment Network; 2010 June 16. www.guardian.co.uk
CDM Watch. Waste Energy Projects. 2010. www.cdm-watch.org
CDM INVESTMENT
Kossoy A. 2010. op. cit.
NORTH AMERICAN CAP AND TRADE INITIATIVES
Western Climate Initiative www.westernclimateinitiative.org [Accessed 2010 Nov 3].
Midwestern Greenhouse Gas Reduction Accord www.midwesternaccord.org [Accessed 2010 Nov 3].
Regional Greenhouse Gas Initiative www.rggi.org [Accessed 2010 Nov 3].
Bianco N, Litz F. Old roads to a new destination. World Resources Institutes; 2011 April 29. www.wri.org
EU EMISSIONS TRADING SCHEME
GLOBAL EMISSIONS MARKET
Kossoy A. 2010. op. cit.

34 Financing the Response
Morgan J et al. Reflections on the Cancun Agreements. Washington: World Resources Institute; 2011.
Mungcal I. Donors missed 2010 aid targets – OECD. Devex. 2011 April 6. www.devex.com
The 0.7% target: An in-depth look. www.unmillenniumproject.org
www.ClimateFundsUpdate.org
STATUS OF CLIMATE FUNDS
A DROP IN THE OCEAN
www.ClimateFundsUpdate.org. Updated April 2011, accessed 2011 April 14). Development assistance includes Official Development Assistance (ODA), Other Official Flows (OOF) and Private, as defined by OECD statistics. The total is based on annual data for 2003 to 2009 and estimates for 2010 and first quarter of 2011, at current prices. From the OECD Query Wizard for International Development Assistance: http://stats.oecd.org/qwids.
Data are provided for 24 climate funds by the Heinrich Böll Stiftung and Overseas Development Institute in the Climate Funds Update web site: www.ClimateFundsUpdate.org. The funds included are Adaptation Fund Board (Adaptation Fund), regional development banks (Congo Basin Forest Fund, Amazon Fund), European Commission (Global Energy Efficiency and Renewable Energy Fund, Global Climate Change Alliance), bilateral donors (Australia, Germany, Japan, United Kingdom, Indonesia), Global Environment Facility (GEF Trust Fund, Least Developed Countries Fund, Special Climate Change Fund, Strategic Priority on Adaptation), World Bank (Clean Technology Fund, Forest Carbon Partnership Facility, Forest Investment Program, Pilot Program for Climate Resilience, Scaling-Up Renewable Energy Program for Low Income Countries, Strategic Climate Fund) and UNDP (MDG Achievement Fund – Environment and

Climate Change thematic window, UN-REDD Programme). The funds have various start dates: the earliest disbursement was in 2003 and the data are up-to-date to April 2011.

CLIMATE FINANCE

Projects approved are from www.ClimateFundsUpdate.org (updated in April 2011 and accessed 14 April 2011). The mapped data include only 24 climate funds overall, and exclude regional or global projects and thus are an underestimate of the climate funding as a total. The bar and pie charts do include regional and global projects.

UK finance

Harvey F. Budget 2011: Green bank is coalition's biggest environmental test. 2011 March 23. www.guardian.co.uk
UK fast start climate change finance. DIFID. www.dfid.gov.uk

World Bank

World Business Council for Sustainable Development (WBCSD). 2010 Nov 30. www.wbcsd.org
www.thegef.org

Adaptation Fund Board

Climate Funds Update
www.climatefundsupdate.org
A complete list of projects and their status as endorsed, approved or rejected see the Germanwatch Adaptation Fund project tracker. www.germanwatch.org/klima/afpt.htm

Hatoyama Initiative

Climate Funds Update
www.climatefundsupdate.org

CLEAN DEVELOPMENT MECHANISM

http://cdm.unfccc.int [Updated 2011 April 15, accessed 2011 April 17]

Part 7: Committing to Solutions

"We live in some of the most..."
Steiner A, Opening of 26th Session of the UNEP Governing Council/Global Ministerial Environment Forum; 2011 February 21 www.unep.org

35 Personal Action

TRANSPORTATION
Gardner GT, Stern PC. The short list: The most effective actions US households can take to curb climate change Environment 2008; 50 (5):12-24.

ENGAGE GOVERNMENT, COMMUNITIES AND COMPANIES
350 Earth http://earth.350.org/

36 Public Action

Toyota reduced...
CORPORATE ENERGY REDUCTION SAVES MONEY
Prindle WR. From shop floor to top floor: Best business practices in energy efficiency. Pew Center on Global Climate Change; 2010 April. p.13.

ASSESSING THE RISK TO BUSINESS
Sussman FG, Freed JR. Adapting to climate change: A business approach. 2008 April. www.pewclimate.org

PARTICIPATION IN POLICY MAKING
Note that Conference of Parties who adopted the Kyoto Protocol became the Members of the Protocol, and sessions from COP11 (2005) onward include both COP and MOP meetings.
Civil society and the climate change process. Cumulative admission of observer organizations COP 1-16. http://unfccc.int/parties_and_observers/ngo/items/3667.php
Virtual participation available through: http://maindb.unfccc.int/public/ngo

http://unfccc.int/meetings/cop_16/virtual_participation/items/5780.php

Part 8: Climate Change Data

"Ensuring the integrity..."
National Academy of Sciences. Ensuring the integrity, accessibility, and stewardship of research data in the digital age. Committee on Science, Engineering, and Public Policy; 2009.

Table

1 UN Population Division. World population prospects: the 2008 revision.
2 & 3 Human Development Index hdr.undp.org
4 FAO Aquastat. www.fao.org [Accessed 2011 May 1]
5 CIESIN. Low Elevation Coastal Zone (LECZ) Urban-Rural Estimates, Global Rural-Urban Mapping Project (GRUMP), SEDAC, Columbia University; 2010. sedac.ciesin.columbia.edu [Accessed 2010 May 28].
Population Division of the Department of Economic and Social Affairs of the United Nations Secretariat, World Urbanization Prospects: the 2009 Revision. www.un.org/esa/population [Accessed 2010 May 31].
6 Sea-level rise: DIVA runs by the University of Southampton for a medium scenario of sea level rise: A1B medium regionalized without adaptation, number of people flooded by country. Courtesy of Robert Nicholls and Sally Brown.
7 & 9 International Energy Agency. Paris, France. www.iea.org
8 Climate Analysis Indicators Tool (CAIT) Version 8.0. Washington, DC: World Resources Institute; 2011.

Photo Credits

The publishers are grateful to the following for permission to reproduce their photographs:

9 Wangari Maathai: Martin Rowe; Philippe Cousteau: Animal Planet; 20 Larsen B ice shelf, NASA; 22 California fire: Gabriel Schroer; Russia fire: Evgeny Prokofyev; China floods: pkujiahe; Australia floods: Andesign101; 24 Larsen B ice shelf, European Space Agency; 26 South Cascades Glacier: USGS; 27 Perito Merino: Sebastien Cote/ iStockphoto; Monteratsch Glacier: www.swisseduc.ch; Glacial lakes, Bhutan: Jeffrey Kargel, USGS/NASA JPL/AGU; 28 Coral: Healthy: Martin Strmko / iStockphoto; Bleached: Rainervon Brandis / iStockphoto; 29 Coccolithophorids: J Cubillos; 32 Isabelle Lewis, based on Reto Stöckli / NASA Goddard Space Flight Center Image; House tipped over: Tony Weyiouanna / Shishmaref Relocation Coalition; Sudan: Claudia Dewald / iStockphoto; Sundarbans: Krishnendu Bose of Earthcare

Films; 44 Pgiam/iStockphoto ; 50 Methane hydrates: NOAA; 54 Supermarket USDA/ David F. Warren; 58 MDBA / Irene Dowdy; 60 Adelie Penguins: Jan Roode / iStockphoto; 61 Bengal Tiger: Stefan Ekernas / iStockphoto; Banksia: Clown and the King / iStockphoto; 67 Flood: WHO; 68 wetlands: brytta / iStockphoto; 69 Tuvalu: Mark Lynas; 70 Vancouver: Ryan Lindsay / iStockphoto; 72 Arctic: Anouk Stricher / iStockphoto; Thoreau's Woods: Jordan Ayan / iStockphoto; Chan Chan: David Rock / iStockphoto; Maple syrup: ImageInnovation / iStockphoto; 73 Venice: RelaxPhoto.de / iStockphoto; Mt Everest: Peter Hazlett / iStockphoto; Thailand: Ine Beerten / iStockphoto; 74 Craig Hill / iStockphoto;76 Colombia: Carmen Lacambra; 77 Cambodia: © Nestlé / Sam Faulkner; Australia: MDBC / Irene Dowdy; Kenya: Green Belt Movement International; 78 Climate champions: British Council; Coffee picker: Rainforest Alliance; Bangladesh: Bangladesh Ministry

of Environment and Forests; 81 Solar Water heater: www.carbonprogrammes.co.za; 83 Small hydro: World Bank / Dominic Sansoni; Wind: Jason Stitt/ iStockphoto/; Biomass: NREL / Warren Gretz; Geothermal: NREL / Joel Renner, INEEL; Solar: NREL / Robb Williamson; Tide, wave, ocean: Sustainable Energy Forum 86 Tree: Adrin Shamsudin; Satellite image: eros.usgs.gov/imagegallery; Conservation worker: UNDP / Chansok Lay; 86 UN Climate Talks; 88 International Policy & Action UN Climate Talks; 100 Mohammed Moradalizadeh / iStockphoto; 102 Window sealing: Cameron Whitman / iStockphoto ; Thermal insulation: Alena Brozova / iStockphoto; Tap: Vasiliki Varvaki / iStockphoto; Solar panels: NREL / Rob Williamsonl 103: Elephant: Daniel Dancer / 350Earth.org; 105 Demonstrators: Global Images; Carbon counter: Deutsche Bank; 106 Felix Alim / iStockphoto.

Index